Group Activities for
Families in
Recovery

*To our families, especially Sam, Jessica, Ky, Miah, Ian, Adam, and Eli;
to the families we work with and learn from; and to our
wonderful coworkers, interns, and teachers.*

Group Activities for
Families in
Recovery

Joan Zimmerman
Jon L. Winek

Appalachian State University

Los Angeles | London | New Delhi
Singapore | Washington DC

Los Angeles | London | New Delhi
Singapore | Washington DC

FOR INFORMATION:

SAGE Publications, Inc.
2455 Teller Road
Thousand Oaks, California 91320
E-mail: order@sagepub.com

SAGE Publications Ltd.
1 Oliver's Yard
55 City Road
London EC1Y 1SP
United Kingdom

SAGE Publications India Pvt. Ltd.
B 1/I 1 Mohan Cooperative Industrial Area
Mathura Road, New Delhi 110 044
India

SAGE Publications Asia-Pacific Pte. Ltd.
3 Church Street
#10-04 Samsung Hub
Singapore 049483

Acquisitions Editor: Kassie Graves
Editorial Assistant: Elizabeth Luizzi
Production Editor: Brittany Bauhaus
Copy Editor: Kim Husband
Typesetter: C&M Digitals (P) Ltd.
Proofreader: Eleni-Maria Georgiou
Cover Designer: Candice Harman
Marketing Manager: Lisa Sheldon Brown
Permissions Editor: Adele Hutchinson

Printed in the United States of America

A catalog record of this book is available from the Library of Congress.

9781452217932

This book is printed on acid-free paper.

12 13 14 15 16 10 9 8 7 6 5 4 3 2 1

Contents

Part 1

Introduction

The purpose of this source book is to provide structured activities for families that are involved in substance abuse treatment.

When considering substance abuse treatment, we must consider contextual issues. First, we must be concerned with the context in which the substance abuse takes place. An important component of the development and maintenance of substance abuse is that people with addiction live in families that also have significant issues. This is not to blame the family for the addict's difficulties or blame the addict for the family issues. Both co-occur, and both need to change for lasting change in addicts and their families.

Second, we must consider the context in which treatment takes place. In our experience, removing the addict from the family for treatment and then returning him or her to the family after "successful" treatment is a recipe for relapse. Rather, the whole family (not just the addict) needs to make significant changes in its structure and process, creating an environment that will allow the recovering addict to maintain sobriety. Treating families as a whole presents some unique issues and challenges for the service delivery system.

While family is an important issue in substance abuse treatment, work is also needed to support the addict in getting clean, staying clean, and preventing and dealing with relapse. The activities included here are designed for the recovery portion of treatment that follows successful detoxification from substances. Detoxification for some drugs can cause severe adverse physical and emotional symptoms and, as a result, should only occur under the supervision of trained personnel. Our family-based activities for families in recovery is part of an intensive outpatient program (IOP). The IOP treatment program for substance abuse is a 3-hour group that meets three times/week. In our group, we use the Matrix model materials (Center for Substance Abuse Treatment, 2006) during two meetings each week, addressing individually focused issues related to substance abuse and recovery. In the third meeting of the week, we invite family members and supports to the group and facilitate these family-focused activities. The assumption is that because substance abuse impacts the entire family, effective treatment requires both individual and family treatment.

WHOLE AND PARTS

Most of the exercises in this workbook came out of our work with a family-oriented outpatient substance abuse treatment program (Winek et al., 2010). In this program, the substance abuser meets in individual group twice a week, and once a week, the group meets with families. The substance abuser also has a counselor that works with him or her (and the families when appropriate) and the person's substance abuse is monitored via urine analysis. Thus, this workbook is part of a larger whole. It is designed to be utilized in its entirety in situations in which the leaders guide the group through a different exercise at each meeting. It is also appropriate to use it in parts, where the leader selects exercises that are relevant for issues as they arise in the group. Our group often gives rise to issues that clients are invited to process in their individual work. At times, clients also bring issues from their individual sessions to the family group. Thus, we see our work with the group as both a whole and a part.

The exercises could be implemented as part of a larger treatment program, as we use them, or as stand-alone treatment. If they are utilized as a stand-alone program, it would be important to screen for clients with severe addiction issues. It is essential that the level of care available to the client is appropriate to the client's needs. If it becomes apparent that the addicted person needs additional support or therapy, it becomes the responsibility of the group leader to assist the client in getting this support. Also, it is important that the group leaders are available briefly after group to assist clients that are triggered by the group process. Although this produces a time burden on the group leader, spending half an hour after a group may be more effective than helping a client recover from a relapse.

ORIGINS

The activities in this book have been utilized in the family group described above over an extended period of time. Most of the activities have been modified after being utilized in group, and many have been modified multiple times. Like most group leaders, we beg, borrow, and share ideas among other leaders and sourcebooks. We have tried to be diligent in tracing back to their origins the ideas we base our exercises on. In some instances, however, components of our exercises have been in common usage for so long that we have been unable to give full credit where it is due. It is a short journey for a really good idea to become part of the treatment culture, and in these instances, the origins are often quickly lost. Thus, we apologize in advance if we fail to give full credit to some of the ideas we build on and welcome feedback about the origins of these ideas.

ASSUMPTIONS

In our view, it is important to spell out the assumptions that our exercises are based upon. Speaking our assumptions that are embedded in our approach allows the reader direct access to the foundations of our work. Rather than make the reader search the text for clues to our assumptions, we chose to simply state them.

At its very core, the exercises in this book are based on systems theory. Systems theory is a theory about how parts of ecology are connected to form a whole and how the whole (in this case) gives rise to an addiction. Once an addiction is established, we become concerned with how the ecology of the addict unwittingly participates in the maintenance of the addiction. Thus, on the most basic level, we can say that given the addict's family situation, biological makeup, experiences, and history, the addict engages in the "natural" amount of drug or alcohol use. From here, it follows that the ecology needs to change in order for abstinence to be the natural state of the family ecology.

Our approach to working with families with substance abuse is based on a contextual world view. This perspective is a little difficult to grasp at first because it inverts conventional wisdom. In the conventional world view, we see clients as mostly (if not wholly) autonomous from the social and emotional context. Our worldview is built upon an emphasis of the importance of the context that the clients live their life in. We have found, however, that this view allows us to understand addiction within the context in which it occurs. It also advantages the therapist by providing a deep understanding of addiction.

Our position rejects theories that see addiction as the result of a single cause. For example, the view of the addict as a moral failure, someone with a genetic predisposition, or the product of a multigenerational process is seen as overly simplistic. Rather, given the whole of the person's life experience, his or her living situation, history, biological predisposition to addiction, and significant relationships, that person is doing what his or her environment mandates. Given this, it becomes clear that it is not enough for the individual to just stop drinking. If the addicted person simply wills him- or herself to stop using, the environment that helped give genesis to the use is still intact.

Thus, a contextual perspective focuses on multiple issues that almost always coexist in the addict's life and family. Given that addiction occurs in the context of the environment, seeing the context as a focus of treatment creates the opportunity to affect the functioning of the addict's family system, thereby affecting the functioning of the addict.

When we say a "natural" amount of use, we simply mean natural in terms of what the environment supports. Given this, we see that treatment that does not pay attention to the environment is, very often, doomed to failure. Often, addicts enter a period of sobriety and clean living that is short lived if the environment is unaltered. The addict often slides back into old behaviors (including using) that are supported by that environment. This can often be avoided by working with the whole of the family system, creating an opportunity for change in the environment, creating a context that supports sobriety rather than addictive living.

EVERYONE IS IMPACTED

Given the relational emphasis of systems theory and our clinical experience, we see that everyone in a family is impacted by substance abuse. Impacts can be direct or indirect, but the lives of the entire family are impacted by the addict's substance abuse. Thus, it is extremely important to work with whole families (or as many family members as possible). Given this, we are frequently encouraging and welcoming of extended family members. We frequently invite clients to bring nonrelated parties who provide our clients emotional support and accountability. For our clients, emotional distance and being cut off from families is common. Although many of our clients have some significant family-like relations they can call on in a time of need, most of them reach us after years of living in their addiction, generally resulting in profound wounds to relationship with their families.

One important way healing takes place in our groups is by allowing an opportunity for family members to talk with each other about how they are impacted by the substance abuse. Alcoholic/addictive families often have unspoken rules of secrecy, and many of the families in our groups have not talked about these issues before coming to group. All families have privacy rules, but alcoholic families take privacy too far and develop rules of secrecy. In our minds, this secrecy is a darkness that allows substance abuse to flourish. Talking with each other (both with their own family members and with other families who have had similar experiences) about the impact of substances on their life brings light to the darkness of secrecy.

LEARN FROM LISTENING AND WATCHING

Part of systems theory is its emphasis on treating the patterns of interaction that support the family's problem. Since these patterns occur on an abstract level, it is often difficult to observe them in your own family. Likewise, the secrecy that is often present in families that struggle with addiction leads to an assumption that the problems of your family are unique. This is especially salient for the children who are raised in families with addiction issues. However, when you are in group with other families and you observe a pattern in someone else's family, you can often become aware of the pattern in your own family more readily. This also helps the families realize that they are not the first family to face an addiction. Thus, our families learn about themselves by seeing other families struggle with similar issues.

Families can also utilize what they learn about other families to develop new ways of coping within their own family. We have heard many of our group members, both addicts and family members, share their feelings of intense relief at realizing they are not alone with their experiences.

CHILD FOCUS

As we stated above, addicts often grow up in families in which their parent or parents struggled with addiction. This lead to an assumption of our approach that holds that addiction is not only a family problem but also a multigenerational family problem. We strive to be mindful of the impact of the addiction on the families' children. Our exercises have modifications for when children are present, and we think that having children present leads to the best outcome. We also strive to be ever mindful of the impact of the addiction on the children. In part, we hold the focus on the children by asking ourselves and each other, "How does this impact the children?"

HEALTHY CARING, NOT ENABLING

Enabling is a process that is common in substance-abusing families: One person enables the substance abuser by blocking the addict from experiencing the consequences of addictive behavior. Typically, the person who acts as an enabler gets her or his sense of self from others. This is very common in our society and, in fact, in many situations is idealized and taught to others.

We find that it is helpful to differentiate between healthy caring and enabling. We have discussed elsewhere (Winek et al., 2010) that enabling is often only perceived after we have crossed over from healthy caring. In its simplest form, enabling is continuing to support a person who does not appreciate the support or who utilizes the resources to unhealthy means. For example, healthy caring may involve paying your adult child's rent during a period of transition, such as after the birth of your grandchild. However, paying your child's rent when he or she is using other funds to purchase drugs would cross over into enabling.

We find that family members often struggle with and respond negatively to the concept of enabling but are more open to the idea of "healthy helping." Coming from the perspective of learning how to help their loved one in a way that is helpful in the long term seems, in our experience, to be an almost universally desired goal for family members. Processing the difference between healthy caring and enabling is an important issue for the group to discuss. By talking with others with similar issues, the clients are more likely to find a healthy balance between supporting and enabling.

FAMILIES AND SUBSTANCE ABUSE—A BRIEF OVERVIEW

As stated above, family therapy concerns itself with the patterns of interaction that occur in a family. On a basic level, the interactions that go on between family members are referred to as *family process*. Family process involves how family members communicate and interact with each other. We can contrast this to "content," which is what families communicate about. Family therapy involves working with the family's process to help the family develop new processes, or ways of interacting. So a family therapist helps a family move from an unhealthy process that gives rise to and supports addiction to a more healthy family process that supports sobriety.

In developing a systemic understating of family systems, one has to consider the relationship between process and structure. Structures are enduring patterns and rules of interaction that occur over time. For example, parents working together to rear the children are a healthy structure. *Structure* refers to the family's process over time. When processes are repeated day in and day out, they become fixed and patterned. Structural family therapy developed by Minuchin (1974) and his colleagues is a highly successful and validated approach (Winek, 2010) that seeks to actively alter structure in families to help families develop more healthy structures and, ultimately, healthier functioning. In many of the activities in this workbook, the desired outcome is an exploration of the family's past and current structure and an invitation to identify and develop more healthy structures. Let us explore some key family structure concepts below.

ROLES

Role theory has a long history in sociology (Mead, 1934; Merton, 1949). In this approach, we consider the rights, duties, expectations, and norms in each social role we perform. From this perspective of role theory, people evaluate themselves and others based upon their ability and performance of their various social roles. Examples of social roles in families include parent, child, provider, partner, and so forth. We speak of role expectation as what is expected by someone in that role. For example, we have an expectation that the role of a parent is to keep his or her young child safe. Role strain is the effort required to perform a role, while role stress is the stress we experience as we perform our various roles.

Role theory is an established social theory that has been applied to a wide variety of situations. In family and substance abuse, it is a particularly robust way to view the ways families function. Role theory looks at social position as a series of roles that people perform in order to accomplish the goal of the organization. In families, we identify basic goals, which are to provide food, shelter, and clothing for family members. Families with children have the additional goal of providing nurturing and socialization necessary for the development of the children. In order to accomplish these goals successfully, family members organize and divide their labor.

Dividing labor is necessary in order to accomplish the goals of the family. Historically, roles were divided along strict gender lines and were prescriptive in nature. Challenging the gender basis of the roles was an accomplishment of the feminist movement. Despite increased flexibility in how families assign roles, however, families still need to divide labor. In healthy families, roles are flexible and based upon skills and talents. For example, Jon is a good cook and enjoys cooking, while his wife does not. As a result, he cooks most of the meals for his family. However, when he has to work late, Jessica is able and willing to provide a healthy meal for the family. In families, roles are both assigned and volunteered for. A common problem in addicted families is that roles are often assigned to family members who care the most or have the most anxiety. This leads to a situation in which children become parentified by taking on roles of the addicted parent(s).

In families with addiction, roles often become rigid and prescriptive. "Only a woman can cook a meal" or "only men can work for pay" are examples of rigid and prescriptive roles. In addition, over time, as the family members become more addicted, they often stop performing specific roles. For example, it is not uncommon for alcoholic parents to stop preparing meals. The children are left to fend for themselves and raid the cupboard. Candy for dinner is not an unheard-of adaptation that children will move toward when their parents are too inebriated to prepare a meal or supervise their food choices.

Roles also need to be developmentally appropriate in terms of difficulty and number. Simply stated, younger children should have fewer and simpler roles than older children. In healthy families, members are supported and trained to be successful in their role performance. Unhealthy families tend to assign roles that are not appropriate. For example, older children are often assigned the role of caring for a younger sibling before the older children are ready for that responsibility. Helping families negotiate more healthy roles is an important function of family groups.

RULES

Rules refer to the negotiated patterns of interaction that family members utilize to govern their interactions. Rules establish boundaries as well as limits to behavior. For example, looking at each other when we speak is a rule that promotes good communication. However, as a family becomes dysfunctional, it often develops unhealthy rules. Just as or perhaps even more harmful, families can fail to develop rules, resulting in chaos. A typical pattern is the addicted family developing rules mandating secrecy, prohibiting family members from talking about what is going on in the family. Families with addiction often develop rules forbidding the children from speaking their feelings about their parent's addiction and the family's problems, which can lead to further inhibition around talking about one's feelings.

As a family develops, the definition of what healthy rules are often shifts. For example, the rules parents set for a child will change as the developmental needs of a child change. Some families continue to keep rules longer than they are appropriate. This can result in developmental delays, where rules in fact inhibit a person's normal development. For example, to protect a teen, parents do not allow her or him to date an age-appropriate peer. This rule against dating can prevent the teen from developing healthy dating skills. In order for a family's rules to change as they need to, the family needs to have the flexibility and ability to negotiate new rules. When a family member has an active addiction, particularly when the addict is a parent, this can be very difficult. Several of the activities in this book invite the family to discuss their rules, evaluate the rules' current usefulness, and work toward developing new ones.

BOUNDARIES

Boundaries were a key theoretical component of structural family therapy (Minuchin, 1974). A boundary is a barrier that separates an individual or system from others and from the environment. In families with an addicted member, there are often loose or inappropriate boundaries. For example, it is not uncommon for an addicted family to let drug addicts and people who engage in criminal activity into a home where children are present. Healthy families have boundaries that allow good influences to enter their family space and maintain boundaries to protect family members from people and situations that are unhealthy. Several of the exercises target assisting client families to establish and maintain more appropriate boundaries.

There is also a need for boundaries between generations of a family. It is frequently the case that children have been parentified. They are often treated as coparent when an addict parent becomes absent. Such a situation puts stress on the child, who suffers undue anxiety as a result. It is important to help the family establish appropriate boundaries between the generations so children can accomplish age-appropriate development tasks.

FAMILY ESPRIT DE CORPS (COLLECTIVE SENSE OF SELF)

Esprit de corps or family identity is the collective sense of self that exists in a family, similar to the personality or identity of an individual. Esprit de corps is the identity or personality of the collective we call "family." Like the negative self-identity an addicted individual can have, a negative family esprit de corps can have a profound and hurtful impact on the family.

The esprit de corps of a family is passed down through generations in the form of statements or concepts that are generalizations about the family. For example, an esprit de corps a client might have about his/her family is that "members of our family don't finish school." Another example is that "we help each other when we need help."

Less healthy examples of esprit de corps tend to communicate an acceptance of negative traits. We must realize that changing an esprit de corps can be difficult, as clients often feel that if they live outside of that mandate, they will lose their family identity.

In substance-abusing families, a particular difficulty around the family esprit de corps can be in the common elements of denial and secrecy. This increases when family members have the added social stigma of being involved with illegal activities. If members talk about issues around illegal actives, the family member can face legal consequences. This inhibition of talking about issues can impact group process and inhibit the development of a healthier esprit de corps. Several of the activities in this book revolve around helping the families articulate and move toward a more healthy collective sense of themselves.

Part 2

Structure of Family Group for Families With Substance Abuse

GROUP PROCESS AND STRUCTURE

In order to develop a systemic approach to family group, it is important to consider both the process and structure of the group. The process of the group address such issues as leadership, avoiding resistance, coleadership, working with diverse families, avoiding resistance, and locus of control.

How the group discusses issues is of great importance as we empower clients to take control of their lives and overcome their addictions. Like family process, group process is how group members communicate with each other and is often expressed in patterns of interaction. In theories of group process, the most important factor is group leadership. However, structure is of equal importance when considering family group treatment for substance abuse, and in practice, they are different sides to a whole. When considering issues of group structure, we consider such issues as recruitment, dual diagnosis, length of sessions, rules, and screening.

If one observes how one speaks about oneself and one's issues, we can get insight into the person's process. Over time, the process that clients have in their lives becomes similar to the way a therapy group will evolve. In systems theory, this process is called *isomorphism*. Isomorphism is the tendency of systems to replicate themselves on different levels. This can occur on many levels; for example, a family that has a depressed member might present as hopeless and helpless about being able to help their member. This feeling of depression as expressed in hopelessness and helpless might be transmitted to the group leaders, who, after a session, feel ineffective and hopeless that their work is helping anyone. The isomorphic process can continue into supervision, in which the therapists working with the client present the case to the supervisor in a hopeless manner.

It is important to note that this process is natural and often unavoidable. We find that the best way to address isomorphism when it presents is to address it in yourself. In the example above, the therapist would be best served by focusing on her or his feelings of hopelessness and helplessness. Resolving your issues makes you more available to be supportive as you encourage the clients to confront their own feelings.

PROCESS

Like family process, group process is who says what to whom in what way with what outcome. Simply stated, it is what happens during group. It is the group leader's or leaders' responsibility for the group to have a positive effect on the members. In order for the group to have a positive effect, the group must be healthy. In order for the group to be healthy, members have to be safe and respected and heard.

GROUP LEADERSHIP

Much has been written about the qualities of successful group leaders, and it is not our purpose to repeat that literature. Instead, we will focus on some unique aspects of family group leaders. To us, the most important common factor with a successful family group leader is curiosity about how the family functions and how it could function better. By being curious, the leader invites the client to consider how things could be different in the client's family. Curious leaders also take a position of leadership that is next to not above the clients. This position avoids a good deal of resistance that is so typical in working with addicts.

Successful family group leaders also display a good deal of sensitivity to a variety of issues. They need to be open to the stories about the clients' struggles in a nonjudgmental way. This allows the clients to learn from their history and move forward toward more healthy functioning. Leaders develop sensitivity to clients' nonverbal behavior and support clients as they learn to speak what has often gone unspoken for so long.

In addition to the qualities of successful leaders, there are some qualities associated with poor success in the group. Group leaders who focus on getting the client to behave "right" are often frustrated by the client's expression of free will and oppositional defiance. This is a fear-based position in which the therapist is afraid of the client's behavior and lack of success. From here, it is a short trip to the unfortunate situation in which the therapists are working harder and are more motivated than their clients. We also find that thinking about therapeutic intervention from a "right" or "wrong" perspective is simply unproductive. It parallels the black-and-white thinking that is often present in an addict's thinking. This style of thinking is often indicative of a lack of autonomy on the part of the therapist. At the same time, this stance is sometimes difficult to avoid, and successful group leaders lean on and are willing to take feedback from (they ask for feedback) their cofacilitators to help sidestep this and other pitfalls.

AVOIDING RESISTANCE

To us, the hallmark of a successful group leader is the ability to access the motivation of the client. We think that resistance on the part of the client is simply the result of the client being unmotivated. Using terms like *lazy, nonresponsive,* and *resistant,* in our view, is simply name calling and is not helpful. The best way to avoid situations in which the therapist is frustrated by the client's lack of progress is to locate what motivates the client. That is, find out from the client's perspective what would drive him or her toward a healthier lifestyle. Having the family assist the client in this search is helpful. In particular, helping clients who are parents to be mindful of their child's need and desire for a sober parent can be a powerful motivator.

Another way to help motivate the client is to be sure that the therapists and other group members are not working harder than the client. Some encouragement is appropriate, but when the group and its leaders are giving a pep talk and/or talks about living up to potential, they are working harder than the client. Paradoxically, as a group grieves a member's failure and moves forward, the client is invited into ownership and is more likely to take responsibility for her or his own behavior. With responsibility comes ownership and motivation to work on one's addiction.

COLEADERSHIP

We have found that leading this type of family group requires coleadership. The groups can get quite large, and having at least a second set of eyes is most helpful in observing the needs of the group. Also, two leaders are well suited to de-escalate any angry and hostile client. If one member runs the group, it is often too easy for him or her to become drawn too far into the drama that is going on in the clients' lives. A coleader reminds the other therapist which system he or she belongs to when the isomorphic process becomes emotionally powerful. Successful group leaders are skilled at working as a team with the cofacilitator, sometimes speaking directly with the cofacilitator during group about group process.

STRUCTURE

Structure refers to enduring qualities of a group. The first issue to consider in terms of structure is group membership. It is our experience that the more clients bring their family members to group, the more dynamic the group will likely be. However, if members are not able to bring family members, that is acceptable as well. Families often emotionally cut off from addicts while the addicts are using, only to reconnect during periods of sobriety. If a client is not in contact with his or her family at the start of treatment, he/she is likely to reconnect at some point in recovery. As he or she reconnects, it would be an appropriate time to recruit the family to participate in the family group.

Also, our definition of "family" is broad. Group members are welcome to bring nonrelated supports. Any person the addict finds as a significant support would be welcome in family group.

RECRUITMENT

Recruitment is the process of engaging family members in the therapeutic process. The idea of treating the whole family for substance abuse runs counter to conventional thinking, which makes recruitment somewhat difficult. As a result, the therapist needs to be proactive in his engagement of families. If you are passive in your approach and simply expect the families to show up, you are likely to fail. We have found that several principles assist the therapist in being successful in family recruitment.

Early engagement is a key principle to engaging families in treatment. During the intake session, it is useful to get a release of information for all members of the client's family. When we request the release, we simply explain that we are a family-based program and we want to talk with all family members. If clients have any hesitation, we simply ask the client a series of questions about the involvement of family in her life. Ask her who knows that she is at the appointment. Who will the client talk with about the session afterward? Who is impacted by her problem? Who is impacted by her success? We sometimes ask who would show up at the ER if the client got in a wreck while using and driving. Following these questions, you can point out that since these people are involved and supportive of her, they would be likely to be willing to support her in therapy. Hopefully at the end of the first session, you have releases for family members and their phone numbers. If this is not the case, don't give up; there will be future opportunities to recruit family members.

This leads to the next principle of direct contact. If you ask a client to recruit his family members, his anxiety will be difficult to overcome. He is either unlikely to ask his family members or, if he asks, he is likely to ask in a way that makes the family members unlikely to be interested in attending. For example, he may invite his family to attend therapy during a fight when the family member is angry with the addict. Experience has taught us that calling the family directly is the most effective way to invite them into therapy. Since cell phones have become so common, it is easy to reach out to the family during the session. If someone is not available, leave a message inviting the person to call you back.

Once you contact the family members, you need to establish an idea of what their agenda for the therapy is. Most family members perceive that the therapist is contacting them to criticize them for the addict's bad behavior. This comes in part from a family's natural feeling of responsibility for each other and in part from the view of addiction as a moral failure. It is important to move past these concerns by finding what the family members are interested in. Once you find each family member's motivation, he or she is more available to attend the family session. In instances in which family members are angry, it is best to be empathic toward their feelings.

The final principle for getting families in is for the therapist to have tenacity. We find that as the client and therapist build trust with each other, the client is more willing to bring in his or her family. Trust builds over time, so asking several times over the course of your work with the client is helpful. When a client refuses our request to bring in family members, a good follow-up question is, "What will tell you it is a good idea, safe, helpful, useful, etc., to bring him or her in?"

DUAL DIAGNOSIS

In recent years, there has been much discussion in the substance abuse field about dual-diagnosis clients. A client with a dual diagnosis is a client who, in addition to a substance abuse diagnosis, has a mental illness diagnosis. Common comorbid illness include mood disorders such as depression, anxiety, and bipolar disease, personality disorders such as borderline personality disorder and sociopathic personality disorder, and developmental disorders such as ADD or ADHD. Some programs place emphasis on treating the substance abuse or the mental disorder before addressing the others. Our system's perspective views the two types of difficulties as related, and we prefer to address both disorders at the same time. We prefer that the client receive therapy for the comorbid disorder concurrently. It is important that the therapist have a release to talk with the client's other therapist to collaborate and coordinate care. In many situations, the family work can also assist with the client managing the comorbid condition. It is often helpful and supportive to hear of other family members who struggle with similar issues. Since most conditions have genetic components, others in the client's family often have similar disorders.

STRUCTURE OF FAMILY GROUP

In our experience, family group takes longer to conduct than a typical hour or hour-and-a-half therapy session. We run our group for 3 hours including breaks (which also fits the intensive outpatient [IOP] service definition requirements in our state). Having the session in the early evening allows family members who work an opportunity to attend and is not too late for students on school nights. While the session is scheduled for 3 hours, it is wise for the therapist to block off at least 3.5 hours. On occasion, holding the group for a few minutes might allow the session to close on a high note. The extra time might also be useful in case a participant needs a bit of extra support after being triggered by an activity in group. While the activities are not designed to be highly provocative, we cannot predict when someone will be emotionally triggered. Having time after the session available might prevent a relapse in a client after a group session.

Since the time of the session is during the dinner hour, on occasion we provide food for the clients. Ordering takeout pizza or some other food (group members will sometimes be willing to bring a covered-dish meal) makes for an opportunity for the group to develop connection to each other as members eat together. The act of eating during group is not only practical but also symbolic. It demonstrates that regardless of what is going on for group members, there is an obligation to meet the basic needs of all members.

The family group is structured in four parts. After each part, there is a 10-minute break to allow for transition. The parts, key activity, and approximate time allotted for each activity are listed below.

Part I—Dinner: 30 minutes

- Dinner is served and group begins at 5:00. The group eats together from 5:00 to approximately 5:20 and then takes a 10-minute break.

Part II—Process: 1 hour

- Chairs are arranged in a circle, and group members are expected to be ready to start the next part of group at 5:30. A group leader introduces the group, welcomes everyone (particularly first-time participants), and reminds the group of basic rules (confidentiality, timeliness, no cell phones, etc.).
- Group leader suggests a question for each person to answer as group members introduce themselves. (Examples of questions: What is your peak emotion, what is something you are grateful for, what is a challenge you faced this week?) Introductions generally include name, drug of choice, and clean time. Group members are encouraged to introduce guests when they bring them. Suggestions for this "opening question" are included with each activity.
- Sometimes it can be helpful, during the introductions, to have each person say whether she would like a chance to process or not and whether she would like feedback from the group or not. This is not necessary when the processing time flows, as it often does in more developed groups, but it can be helpful if the processing time is uncomfortable and does not flow well.
- We remind people that since this is family group, it is an appropriate time to process family issues.
- The floor is opened to those who want to share. Group leaders facilitate as necessary. This is the group's time.

Break: 10 to 15 minutes

Part III—Group Activity: 1 hour

- Group activity, led by group member and/or group leader. Group activity includes processing as appropriate.

Part IV—Closing: 15 minutes

- It is important to have a closing each week so that the group feels some closure before leaving. It is also good to check in and ask if anyone has been triggered during group. Some example of closing are:
  - Circle up and say the Serenity Prayer, though this can be controversial.
  - Circle up and have a moment of silence.

- Circle up and have group members share a quote.
- Circle up and have each person say something he/she got from the group.
- Circle up and have each person give a compliment to the person on his/her right.
- The group will have good ideas about how each would like to end group.

CHILDREN IN GROUP

For us, the issue of including children in group is not *if* they should be included but *how* and *when* they should be included. We provide child care in an adjacent area for younger children. The leaders, if possible working with the parents, consider each child's situation, level of maturity, and personality in deciding in what portions of the group to include the children. Unless the children are very disruptive, they are almost always included in the meal portion of the session. It is at the group leaders' discretion whether the children should be included in the process session. It is likely that younger children will be excluded from this portion to protect them from being exposed to age-inappropriate materials and to not impede the other members. We have a very general rule that children should be 15 years old to participate in processing.

It is in the group activity that children are most likely to be included. Again, this decision is made based on each child's developmental level and needs. Our exercises include guidance on how to modify the activity for sessions in which children are invited to participate. Group members must be reminded when children are present. A member is likely to slip and use some inappropriate language at some point during the session, so he/she is simply reminded that children are present, and if a member is unable to maintain appropriate language, the adult rather than the child would be asked to leave.

The process of including children in some activities and not in others is a parallel to the process of healthy families. All families must consider when children are developmentally ready to participate in certain activities. Likewise, there are certain activities that children should be excluded from with healthy, clear boundaries. How this is handled in group can model for group members how to set boundaries without abandoning the children.

The group we lead includes some families that have children and a majority that do not. Thus, it is difficult for us to ensure a child-appropriate environment, since childless adults are often not in the habit of monitoring their behavior in a way that is appropriate for children. Often, the topics that are processed are not appropriate for children. Because of this, our family group often is an adult family group. In this situation, it is helpful to explain to parents that children are not included in the group to keep the children safe rather than because they are not welcome. Also, parents are often used to including their children in adult activities, and working with parents to identify that pattern as well as discriminate between "adult" topics and processes and those that are appropriate for children can be helpful.

This book is written for both adult-only family groups and groups with children. There are, however, a few exercises that are not suitable for use in groups with young children. These are noted in the exercises.

FAMILY GROUP RULES

All social groups have rules that govern safe, respectful, supportive, and healthy interactions. Rules that are clear and explicit are better than implicit and or unclear rules. It is primarily the responsibility of the group leader to enforce the rules. However, all members have a stake in the following of the rules, and it is expected that the whole group speak up when she or he has a concern about another member of the group. The individuals in our group are involved in an intensive outpatient group program that provides individual-based group treatment twice a week. The rules for those groups also apply to family group. We also have some unique rules for family group:

1. Group members must be on time, but we are more flexible with family members. They can come late if they have to and are allowed to leave early if necessary.

2. We remind the group of the importance of confidentiality each week, though we also point out that we cannot guarantee that all present will honor that.

3. We ask each week if there are new people present and welcome them. We also say that family members are welcome to participate at any time during the group (sometimes it is unclear to family members what they are supposed/allowed to do. Also, we have found that sometimes family members have been told not to talk by the people they are there to support; we try to dispel this idea).

4. We remind everyone of the importance of respecting others present, particularly not talking when someone else is talking.

5. We talk about the time issues and let each group member know that since everyone needs to be able to speak, we need to respect the time available. We inform the group that group leaders may remind group members of the time available if necessary. We also say that sometimes people do need more time, and that is okay.

6. We ask group members to monitor their language (no swearing, etc.), particularly on family night.

7. Cell phones must be turned off.

8. No "war stories."

9. No new sexual relationships between group members (we allow existing couples in our group).

10. Group members are asked to stay to put the room back in order before they leave.

11. Group members are asked to bring a quote to share at the end of group.

12. Group members can volunteer to help create and cofacilitate an activity related to a family topic. This can be an activity from this book, or one they come up with on their own. The topic and activity must be related to family issues and must be approved by a group leader.

It is helpful if members are provided a copy of the rules. As new members join the group, they can be oriented to the rules in the first session. This serves to orient the new members as well as remind the continuing members of the expectations for their behavior. Likewise, the clients are reminded of the confidentiality rule in most sessions.

SCREENING

Screening is the process of determining what clients are appropriate for group. As mentioned above, we do not screen out a client for mental illness as long as the client is able to participate in the group appropriately. Many of our clients are court involved, which is not a problem as long as they sign a release for us to communicate with the court counselor or probation officers. In fact, that is useful leverage to assist the client in his/her decision to attend the group with his/her family.

Screening occurs in sessions prior to attending a group session. In instances in which a client has been accepted in group but is disruptive, continues to relapse, or has a serious violation of the rules, the client will be given a session outside of the group time (often meeting with the whole treatment team) to discuss her/his plan for improvement. On occasion, we need to exclude a client from participating in the group until she/he is better prepared. This most often happens when a client relapses and needs inpatient treatment. In these situations, family group can become an important part of the addict's aftercare upon discharge from inpatient treatment.

Our clients often have a history of violence toward their partners or engage in mutual combative situations. In such situations, we must evaluate whether the person is appropriate for group. What is of concern to us is how the substance use relates to the violence, how recent it was, and whether the client is committed to living a violence-free life. Looking at these safety of both partners factors on a case-by-case basis helps us make a decision about allowing the client to attend the group.

Many of our clients also have been or are currently involved with custody issues after a report has been made to social services. In these instances, it is important that family group therapists coordinate care with social services and the child's therapist when possible. We see having a release and a good working relationship with social services as essential to providing services to that client. Family group is an opportunity for the clients to address their issues as they work to regain and/or maintain custody of the children. We always encourage group members to share their experiences, with the expectation that in doing so, group members will recognize that they are not alone with their experiences. We often hear group members reacting with surprise to this realization.

We will work with dual-diagnosed clients as long as there are not psychotic features associated with their illnesses that become barriers to their recovery and they are able to be in group successfully. We do have concerns with clients that are unable to make a commitment to becoming sober and remaining sober. Clients who are in active addiction that requires detoxification would be referred for inpatient detoxification until they are ready to work with their families in group. On occasion, a client will relapse to the degree that he/she requires detoxification. These cases are addressed individually, but we will sometimes allow the client and his/her family to return to group when detoxification is completed.

PROCESS AND STRUCTURE OF THERAPY

A common perspective is that therapy provides the client with some tangible product. That is, as a result of therapy, the client learns something new or has insight into unconscious process that allows new behavioral options. More experiential and expressive views of therapy see therapy as a process that presents the client with an experience that is transformative. In our experience, both perspectives have merit. Given this, we focus on a "both and" perspective. Often, we are able to provide insight or psycho-education for our clients. Often, we find that the process and structure of the group is the therapeutic effect for our clients. In fact, on occasion, clients are best served by the content of the group, while other clients in the same group seem best served by the environment of the group. We strive to maintain a healthy environment that can promote change in itself while focusing on the content of the group, providing meaningful information and experiences for our clients.

LOCUS OF CONTROL FROM EXTERNAL TO INTERNAL

Where one perceives the locus of control on one's thinking and feelings is an important issue in substance abuse treatment. There is a general pattern that as people develop more healthy senses of self, the perception of the control of their thinking and emotions moves from external to internal. For example, a child that is crying, when asked about his tears, will say, "Billy made me cry. He took my toy." Here, the perception is that the tears are the product of Billy's actions. Contrast that to an adult woman who states that she is crying over the loss of a loved one. Here, the feelings are seen as an internal state that is influenced by external events.

This lack of internal locus of control often gives an addict an excuse to use. The addicted person may rationalize addiction as a response to his or her environment. For example, an addict may state that his use is a way to cope with the stress from family conflict. That is expressed, for example, as, "I drink because my wife is a bitch" or "I use because my husband is cheating on me." Family group presents an opportunity for the therapist and other family members to invite clients into developing an internal sense of control. This is best thought of as extending a series of gentle invitations to adopt this perspective. This occurs by rephrasing a client's statement. For example, "I relapsed because I had a fight with my parents" could be rephrased to "you were so hurt by your argument with your parents that you chose to drink. Is there a better way you could express your feelings about the fight you had, or could you re-engage your parent in a way that would resolve your conflict?" This could be followed by questions about "what would be a better response to your situation?"

Most of our clients come to group with an external locus of control. When asked why they are in group, they often will respond that "Social services is making me come" or "Probation said I have to be here." One of the most concrete shifts we see as group leaders is the transition from this thinking to an internal locus of control, when clients take responsibility for the actions by acknowledging they are in group "because I used." Group members can be helpful in encouraging this transition by confronting statements that put off responsibility on DSS or probation. In general, such confrontations are much more effective coming from another group member rather than the group leader. Group leaders can learn to use their skills to draw out these comments from other group members ("Susan, what do you think [or what do you feel] as you hear Jane say that?") rather than confronting them directly.

CONFIDENTIALITY AND LEGAL AND ETHICAL ISSUES

Compared to individual work, confidentiality is a more difficult issue to address in group and family work. The therapists providing the family group therapy are legally and ethically bound by the principles of confidentiality. However, the members are not bound by such constraints. Given this, the group members have to trust each other to maintain confidentiality. We work to help group members understand the importance of confidentiality and make a commitment to following this principle. We discuss this with clients before they start the group and remind group members frequently of their agreement to maintain the confidentiality of the group.

Frequently, legal issues come up in the group, and we strive to support our clients to live drug free and above the law. When we have clients with legal issues pending, we will support them as they go through the legal system. In fact, the legal system is utilized as leverage, as we point out to clients that a report on their progress from the group leaders can only serve to improve the client's legal case. The other legal issues that come up with some regularity are issues of child endangerment and abuse.

In North Carolina, we are mandated by law **to make** child abuse reports, and we take that responsibility seriously. This is disclosed in our release forms, and we talk about it on an as-needed basis. When a client makes a disclosure that requires us to report, we try to engage the client in that process. We will work with the client and encourage him/her to make the report to social services. We also offer to help him or her make the report. We discuss with the client how self-reporting can improve his/her standing with social services, and we will be present with the client as a report is made. In the rare instance in which clients are unable or unwilling to make the report, we make a report promptly. We do not make a report without telling the parents we are doing so unless we are unable to reach the parents or there is a possibility of danger. We find this approach works to help build trust with our clients, as they know we will do the right thing, and they do not worry that we are making a report behind their backs.

BLENDED FAMILIES

A clear trend in families is an increase rate of divorce, which results in an increase in blended families. There are some important attitudes for family group leaders to adopt to assist blended families. It is important to see blended families as different rather than pathological. They can be functional and adaptive if family members are supportive of each other and provide for the children's development. It is important to recognize structural challenges, especially the challenge of step-parenting, and be able to support the family as this issue is addressed. Finally, the therapist needs to recognize that needs of the children change as a function of their ongoing development.

DIVERSE FAMILIES

A clear direction that society is developing in is an increase in diversity. The days in which we could speak of a homogeneous society are gone. According to 2010 U.S. Census data, almost 28% of the population is non–White. We also see that, according to the most recent census data, the growth rate for Whites was 5.7% in the decade between 2000 and 2010, while the population growth for African Americans was more than double that at 12.3%, and the growth rate for Latinos was an astonishing 43%. Given these facts, it makes sense to expect that family group therapy will be attended by a diverse group of families. We have found that it is the attitude of the group leaders that makes families feel welcome and supported. The attitude that the therapist needs to convey is that she/he is open, understanding, interested, and concerned with the culture of each family that attends the group.

It is important for group leaders to have some basic understanding of the local cultures. However, it is unrealistic to expect that the leader will have adequate knowledge to be prepared for all situations. We have found that being able to talk about your own cultural struggles normalizes the clients' experiences and can be helpful. The group is an opportunity to talk about differences in a sensitive, nonjudgmental manner. This invites clients to a process of self-reflection. Being able to reflect on your situation and how you are responding to your context is a step toward health.

STAFFING

Because our group is part of an intensive outpatient program, we schedule weekly staffing with the treatment team to review progress of group members, address group issues, and exchange information. We also have a shorter staff meeting at the end of each group to allow staff to process their experiences. We find these to be essential pieces of running a successful program, creating a stable and connected treatment team, and providing group leaders the support and care they need.

WORKING WITH EXPRESSIVE ARTS

Many of the activities in the workbook are expressive arts activities. Expressive arts are useful tools in assisting clients to express what they are unable to express in words. Often, our clients have learned to repress rather than express thoughts and feelings. This is, in part, a result of the secrecy that is so common in families that struggle with addiction. Both authors have experience working with expressive arts and have had many experiences working with clients that affirm the power of using expressing arts in individual, family, and group work.

At the same time, using expressive arts can be intimidating. We believe that many of us (group leaders and group members) are convinced early on that we are not artists—that we have no talent and we can't do it. Facing the fears these beliefs carry takes courage. As we ask our clients to make art, we are asking them to face these fears, to be vulnerable.

The healing process does not stand by itself; healing in one area (or hurt in one area) affects change on other levels. "We are conditioned to see life as a series of unrelated events and, as a result, we become event-activated: caught off guard by unanticipated problems, we go into action to counter them without considering their larger context, their interconnectedness" (Nichols & Schwartz, p. 122). An interconnected view can help us expand our way of thinking, allowing us to more easily think "outside the box." How on earth can sitting and drawing, for example, trigger a change in our deepest levels, in how we relate to life? Realizing that change in one area affects change in other areas allows us to open our minds to such possibilities. The question becomes how to be flexible enough as therapists to allow clients to explore their growth in whatever direction works for them and sensitive enough to be able to discriminate between a client's lack of interest in a modality due to her own individuality or an aversion due to layers of pain and hurt.

Making art can help us create a vision of possibilities, a path from here to there. Envisioning something different can be a first step toward creating something different, of creating hope. Satir (1983) speaks of hope as a major component of successful therapy: "People do not dare go from the known to the unknown without hope" (p. 175). Breaking through the feeling of "I can't do that" can lead to the realization that "Wow, I did that!" That experience can be transferred to other areas of our lives. Through making art, we can experience (and help our clients experience) that doing something different, though uncomfortable at first, can lead to new levels of confidence and satisfaction. When we tell our group we are going to draw, we are met with eye-rolls and groans. By the end of the group, the majority of group members, surprised, agree they enjoyed and got something out of the process.

Our clients often need to relearn how to feel good without drugs and alcohol. Art-making can help in this process. If we can help them manage to get out of the way of the intellect, turn off the "I can't do it" tape, and avoid the "This isn't good enough" space, many times they will find that the process of making art, whether it be drawing, dancing, molding clay, or painting, just feels good. The strokes of the brush feel good; the clay feels cold and smooth and supple in our hands. If we can stay out of the intellect, making art is fun. We play. We revel in the moment. We giggle. We get absorbed in the process. We smash it and do it again. We re-experience some of the good things. We experience a part of ourselves that does not always show itself. We are often surprised.

But I (the Group Leader) Can't Draw!

See below (how to talk about expressive arts activities), and apply the suggestions to yourself! Having a group leader acknowledge being nervous and uncertain can be helpful—especially when the group leader is willing to talk about it and try it anyway! None of these activities requires artistic talent; they just require willingness. We often remind group members that the definition of being brave is not having no fear but is being afraid and doing it anyway".

How to Talk About Expressive Arts Activities

- There is no pressure to do it a certain way.
- There is no way to do it wrong.
- If you want to do it in a different way, that's okay.
- You don't have to be a "good" artist.
- You don't have to be able to draw.
- You won't have to share more than you are comfortable with (if you don't want to share what you made, you can just share your experience of making it).
- This is uncomfortable for many people at first, don't worry.
- You can create what you want to create.
- Expressive arts helps "bypass" the intellect and access emotional intelligence—so you don't have to think about it! If you find you are thinking about it, go back to the process.
- This is a chance to have fun.
- It can be helpful to not talk while you are creating.

Part 3

The Curriculum

Curriculum Section I:
Family Structure

Family Roles

1

Activity Title: How Families Look: Roles and Structure

Activity Mode: Worksheet and Expressive Arts (acting)

In families, family members tend to take on identifiable roles. Families that have flexible rather than rigid roles tend to be more healthy. Family members can function in different roles at different times. For example, a child may go through a phase of being the "good child" and may at other times be the "difficult child." Or Mom can be the main disciplinarian during some periods and Dad can be at others. Rigidity of roles varies as families develop. For example, a young child needs concrete expectations that a teen would find constraining.

In families with substance abuse, however, family members' roles are often extremely rigid (Wegscheider, 1981). Individuals in the family cope with the chaos of substance abuse by taking on these roles, which often takes focus away from the substance abuser and reduces the family's expectations that he or she participate.

This activity helps group members identify family roles often seen in families with addiction. Group members will begin to recognize how these roles have impacted them individually and their family as a whole. Most often, these roles are recognizable to group members, and group members are often willing to share their experiences.

The idea that changing the roles we play in our families is possible can be a completely new idea and can inspire hope for the future.

This three-part activity begins with a worksheet that helps group members think about and identify qualities in healthier families and elicit discussion of issues such as communication, boundaries, family identity, and roles. The second part identifies family roles. After this discussion, the group breaks up into smaller groups and acts out scenarios in which individuals depict various roles.

Our experience is that though group members may be hesitant to act out the scenarios at first, they will most often become interested in participating as the activity progresses. Group members who say they are too shy to participate in acting can still participate by directing and contributing thoughts, experiences, and ideas.

OBJECTIVES

- To introduce family members to the concept of roles
- To make families aware that rigid roles are usually found in families with substance abuse
- To encourage group members to think about roles in their family of origin
- To acknowledge that children take on roles in families as a way of surviving and often carry those roles into adulthood, even though they no longer need those roles to survive
- To offer support by normalizing "unhealthy" family structure (no such thing as a perfectly "healthy" family)
- To instill hope in the possibility of changing roles, even if they have been rigid for a very long time

MATERIALS

- How Families Look: Roles and Structure worksheet
- Scenarios
- Pens, pencils

OPENING QUESTION SUGGESTIONS

- As you were growing up, what is one thing that worked well in your family?
- What was one healthy thing you did to "survive" growing up? (Examples: take time by yourself, read, be funny, spend time with friends)
- Who was the funniest person in your family (or most intelligent, creative, respected, hardest worker, etc.)?

METHOD

1. How Families Look: Roles and Structure Worksheet: Part I: More Healthy/Less Healthy Family

Tell the group that today we are going to talk about what makes a "healthier" family and what makes an "unhealthier" family. This is a good time to normalize being raised in an "unhealthier" family and to point out that there is no such thing as a perfect family. It is also important to talk about the role from a developmental perspective. What might be a healthy role at one stage of development might not continue to be healthy as the family develops. For example, doing laundry for your child is appropriate when the child is young but not appropriate as the child becomes capable of doing it him/herself.

a. Hand out worksheet (How Families Look: Roles and Structure).

b. Using the board or large paper, elicit answers from the group about what makes a healthy family. It is fine if the group gives answers from the worksheet.

c. Give the group a few minutes to write down three qualities they have identified that are not already on the list.

d. Next, ask the group to name some qualities that contribute to a more "unhealthy" family.

e. Again, give them a few minutes to write down three qualities that are not already on the list.

f. Ask group members to identify three things they can change to contribute to a healthier family. Ask them to take a few minutes to write down their thoughts, and then ask for volunteers to share what they thought of.

2. How Families Look: Roles and Structure Worksheet: Part II: Family Roles

Give information about "family roles" in the frame of a rigid structure in an alcoholic/addict family. See the Rationale (above) for more information about this topic.

a. Ask group members to identify the roles they played growing up and the roles they play now.

b. Emphasize that everyone plays these roles sometimes, but in alcoholic/addict families, these roles can be rigid.

c. Ask group how many people came from alcoholic/addict families and point out that this is very common; encourage hope for "breaking the cycle" and point out that as they do this, their own children will have very different stories.

d. Ask group members to identify some way they can change that will impact the roles in their family. Point out that as children, we take on roles in order to survive; as adults, we can recognize that these roles may no longer be necessary, and we can change.

3. How Families Look: Roles and Structure Worksheet: Part III: Acting Out Scenarios

a. Divide into groups of six and give each group a copy of their assigned scenario. Have the group assign roles, and give them 10 minutes to practice acting out the scenes. Emphasize they can change the scenarios if they want to make them more relevant.

b. Have the groups perform their skits.

c. Ask group members what it was like to play the roles they played. Could they see similarities/differences with their own families?

GROUP DISCUSSION AND PROMPTS

The above directions include numerous prompts for discussion. Others could be the following:

For How Families Look: Roles and Structure Worksheet (Part I)

- Point out that group members can identify qualities from either their family of origin, their current family, or both. What is similar? What is different?
- What are/were some fun activities you do/did with your family?
- Did you grow up with clear and consistent boundaries and limits from your parents? Do you think this is important?

- Many children in addictive families play the role of the parents. Did this happen in your family?
- "More healthy" family does not mean perfect. There is no such thing as a perfectly healthy family.
- Substance abuse impacts everyone in the family—that's why we have family treatment and encourage group members to bring their families and supports to group.

For Family Roles (Part II)

- Who is familiar with any of these roles?
- What role did you play? How skilled at this role were you?
- What stands out to you about these roles?
- Did/does anyone play more than one role in your family?
- What aspects of this role do you want to stop playing? When?

For Skits (Part III)

- How did it feel to play your role?
- Did the skit bring up any emotions for you?
- What did you want your character to say/do?

AVOIDING PITFALLS

- Be aware that some group members may feel guilt and/or defensiveness about their families. It is important to normalize unhealthy family structure and emphasize that there is no such thing as a 100% healthy families. The important thing is to be willing to look at these issues as they come up and move in a direction of increasing individual and family health.
- It can also be helpful to normalize the difficulty of talking about these issues and frame being willing to do so as a courageous step toward health.
- These skits do not have to be long in order for them to be effective. Some performance anxiety can be alleviated if the group members know that the skit need only be a few minutes long, and they can still be very powerful.
- If there are not enough people in the group, not all the roles need to be taken on; the groups can decide which to use or not use.
- Individuals that are extremely shy can be encouraged to help with the directing of the skits. However, we find that once everyone does get involved in creating the skits, much of the performance anxiety disappears.
- As with other activities, this exercise can trigger strong emotions. The skits can remind group members of their own families of origin, and that could be painful. It is important to be aware of group members' responses and make sure there is adequate time for processing. Always offer to stay after group, if necessary, to speak with anyone that may be triggered. At the same time, it is important to emphasize that talking about these difficult emotions, including being triggered, can be helpful, as individuals can then receive support and suggestions from the group.

CULTURAL CONSIDERATIONS

- Group leaders must be cautious about their definition of "family" and steer away from defining and discussing family in any way that discounts nontraditional families.
- Different ethnic backgrounds will be able to identify different roles in families. Group leaders should elicit these descriptions and ask group members to share their expertise about their cultures.

FOR GROUPS WITH YOUNG CHILDREN

- Parts I and II would generally be appropriate topics for children to be a part of, and group leaders may be able to encourage productive discussion between parents and children about how these roles play out in their families and how they can move toward more healthy functioning.
- Care should be taken that the scenarios in Part III be enacted in ways that are appropriate for the children present. Group leaders can work with parents to alter the scenarios in ways that parents feel would be appropriate for their children. This also gives group leaders a chance to model setting appropriate boundaries with children and helping parents consider what is appropriate for children to be a part of and what is not.

HOW FAMILIES LOOK: ROLES AND STRUCTURE

Part I: More Healthy/Less Healthy Families

"More Healthy" Family

- Clear communication—we say what we mean
- Connected with each other
- Connected socially (with others outside of the family)
- Ability to adapt (not rigid)
- Clear roles (everyone has a job and is able to perform that job)
- Intergenerational boundaries (grandparents are in a supportive role, parents provide structure and support, children learn and develop)
- Quality time together
- Strong family identity (think of a family motto)

What other qualities are important in a "more healthy" family?

1. _____

2. _____

3. _____

4. _____

"Less Healthy" Family

- Poor communication (denial of conflicts, mixed messages)
- Enmeshed (too close) or disengaged (too separate)
- Isolated, little healthy social connection
- Rigid rules/secrets
- Confused roles (not everyone has a job nor can everyone do his or her job in the family.
- Poor intergeneration boundaries (children act as parents, parents as children)
- Family time is poor quality, often conflicted
- Grandparents undermine parents
- Weak family identity
- Enabling behaviors
- Very little or no personal privacy

What are some other qualities found in a "less healthy" family?

1. _____

2. _____

3. _____

4. _____

What are four things I can do to move toward a more healthy family?

1. _____

2. _____

3. _____

4. _____

Part II: Family Roles

The following roles can often be seen in families with addiction (Black, 1987). Note that though this is often the case, all families are different and develop their own ways of coping. Think about these ideas and whether they fit with your own experiences.

1. "Addict": the one "with the problem"—the world tends to revolve around this person

   Who fills this role in your family? What are some examples of ways of acting that are consistent with this role?

   _____

   _____

   _____

   _____

2. Hero—this is the person that always does well, looks good, wants to make the family look good, denies there is a problem

   Who fills this role in your family? What are some examples of ways of acting that are consistent with this role?

   _____

   _____

   _____

   _____

3. Caretaker—helps everyone, especially the addict; tries to keep everyone happy; makes excuses for behaviors and actions, never mentions the problems; good intentions

   Who fills this role in your family? What are some examples of ways of acting that are consistent with this role?

   _____

   _____

   _____

   _____

4. Scapegoat—often looks like "the problem;" acts out, takes attention off the addicted person

Who fills this role in your family? What are some examples of ways of acting that are consistent with this role?

_____

_____

_____

_____

5. Lost child—quiet, reserved, doesn't make problems; gives up self-needs, avoids difficult conversations

Who fills this role in your family? What are some examples of ways of acting that are consistent with this role?

_____

_____

_____

_____

6. Mascot—"the jester"—often brings humor (sometimes hurtful)

Who fills this role in your family? What are some examples of ways of acting that are consistent with this role?

_____

_____

_____

_____

Part III: Role Plays

Scenario 1:

Family of six. Mother calls everyone to dinner; father will not come and stays watching TV and drinking. Rest of the family around the table responds to the situation. At the table, Mom is upset, kids respond in various ways.

Mom—caretakes dad

Hero—talks about successes

Scapegoat—acts up, angry

Lost child—quiet

Mascot—jokes, distracts

Scenario 2:

Family of six. Everyone comes home to find Mom gone. Family assumes Mom has gone out using; Dad and family get dinner ready and spend the evening.

Dad (caretaker)—defends Mom, tries to get everything together

Hero—talks about report card

Scapegoat—in trouble at school

Lost child—quiet

Mascot—makes fun of Mom

2

Impact of Addiction on My Family

Activity Title: Before—During—After

Activity Mode: Expressive Arts (drawing)

RATIONALE

Although most people will acknowledge that substance abuse has impacted their family tremendously, it is difficult to comprehend the breadth of this impact—and the breadth of the impact of recovery on their family. This activity is especially powerful for families that are speaking about this for the first time.

This is an expressive arts activity that can help clients work toward acknowledging the impact their addiction has had on them and their family. It can also create an opportunity to visualize a future for their family that is not organized around substance abuse.

For group members that have family members present, this can present an opportunity for the families to work together and begin to process family issues in a safe environment.

OBJECTIVES

- To help group members (both the addict and family members) develop an understanding of how addiction has impacted their family
- To help group members develop a vision of a future in which the family is not organized around substance abuse
- To encourage and support group members and families as they develop feelings of hope related to their families
- To encourage families to work together and begin to process their experiences
- To create a safe space in which families can share their experiences and feelings with each other and with the group
- To allow family members to hear stories from other addicts and family members that normalize their own experience ("we're not the only ones")

MATERIALS

- Paper (preferably a roll of poster paper, cut into approximately 4- by 5-foot sheets; if not available, then several sheets of regular-sized paper that can be taped together)
- Crayons, markers, pens, pencils
- Magazines to use for collages
- Scissors, glue

OPENING QUESTION SUGGESTIONS

- What is one positive change that you have experienced since you (or your family member) stopped using?
- What is one positive change your family has experienced since the family member stopped using?
- What is one thing you worked to change in the last month (besides stopping using)?

METHOD

In this exercise, group members work either individually, with their families, or in small groups to develop a visual representation of the impact addiction has had on their families. A timeline approach is used to represent times in the family when family structure was not organized around the addiction (families may remember a time when addiction was not active in their families) and a future in which that could be the case.

It is important to note that some individuals may not remember a time in their family not impacted by substance abuse (those that grew up in substance-abusing families) and to normalize this experience as it arises.

As with many expressive arts activities, group members may at first be hesitant to participate. In this situation, it is helpful to point out that there is no "right" or "wrong" way to complete the activity, and group members may use words, drawings, symbols, or other means of expression (see Chapter 3, Working With Expressive Arts).

1. Introduce the idea that addiction is a family disease rather than just an individual disease, and it impacts everyone in the family.

2. Ask for thoughts from the group on that topic. Support group and family members as they share how they have been impacted by addiction in their family.

3. Give instructions for each person (families may choose to work together on one project or do individual projects) to create three pictures.

   a. One picture of their family before addiction

   b. One picture of their family during addiction

   c. One picture of their family after addiction (as far in the future as they want)

4. Let group members know there are no limitations on how pictures can be made or what they need to look like. Many people are nervous with art activities, so emphasize there is no right or wrong; they can use words or pictures; they will be able to decide what they will share or not share.

5. Allow 20 to 30 minutes or until most are done with the project.

6. Gather the group together and ask what the experience was like—was it emotional for anyone? Did anyone learn something they didn't know? Who is going to share this with their family members who aren't here?

7. Ask who would like to share with the group. If members are not comfortable sharing their projects, ask them to share what creating the project was like for them—did it bring up particular thoughts, feelings, memories, and so forth?

GROUP DISCUSSION AND PROMPTS

- What was this experience like for you?
- Did you notice any emotions arising for you?
- What did you learn from this activity?
- How many people thought they would not enjoy this activity and then did enjoy it?
- What was it like for families that worked together on the project?
- Were there any surprises?
- Which section was easiest to create (before, during, or after)?
- Which was the most fun to work on? Which was most difficult? Why?

AVOIDING PITFALLS

- Expressive arts activities can bring up fears and shame for many people. Acknowledging this possibility, normalizing those feelings (many of us have been shamed as children about not being "good" at drawing), and emphasizing that there is no "right" way to do this activity can be helpful. Having lots of magazines and opening the door for collages that do not involve drawing can also diminish fears.
- Some group members may not be able to visualize a future (or a past) without a using family member. If this is the case, group leaders can encourage the group members to "use your imagination rather than your rational mind." Some brief imagination exercises (imagine being at the beach, imagine their house painted a different color—anything imaginative) may help group members' minds shift.
- As with many expressive arts activities, strong motions may come up for group members. At the end of the activity as the group members are processing their experiences, group leaders should ask specifically whether anyone is experiencing strong emotions or a desire to use. Group leaders should always offer to stay after group to meet with individuals who have been triggered.
- Occasionally, group members may refuse to participate in the activity. Group leaders should avoid getting into a power struggle with individuals by framing it as a safety issue for the group members, with an invitation to participate when he/she feels safe. In these situations, group leaders may want to set limits on what nonparticipating folks can do (please don't talk to other group members, for example). Group leaders may also require that everyone participate in some way, but each person can make her or his own decision about how to participate.

CULTURAL CONSIDERATIONS

- Expressive arts activities can often serve to lessen the impact of language barriers.
- This exercise can sometimes elicit stories and descriptions of family experience that can increase group leaders' understanding and appreciation for the cultures of group

members. Expressing this genuine interest and appreciation can be very helpful in encouraging such disclosure.

FOR GROUPS WITH YOUNG CHILDREN

- This activity is generally appropriate for groups with young children and can be a helpful exercise for parents to work on with their children. Parents can gain a greater understanding of how substance use in the family has impacted the child, as well as an idea of what the child sees as a hopeful future.
- This could be an emotional and powerful experience for parents working with their children, as they continue to recognize and take responsibility for the impact their substance has had on their children. Group leaders must be sensitive to this and proactively offer support for parents.

Before-During-After Example

3

How We Experience Our Family

Activity Title: Family Sculpting

Activity Mode: Expressive Arts (experiential—acting)

RATIONALE

In all families, each individual has her or his own unique experience, and individuals struggle to understand each other's experiences. In families with substance abuse, this struggle can be magnified. Often, issues are not openly talked about, acknowledged, or addressed. As the family organizes around the substance abuse, individuals often worry that addressing issues may be too risky or too painful.

The process of families moving into recovery can often involve a deepening understanding of each other's experiences—both for the substance abuser and for family members. Beginning to talk about such issues can be a big step for families, and this exercise can help them further that process.

Sculpting is an experiential activity, developed by Satir (1983), that makes us aware of structure and process in families. Sculpting requires participants to "direct" other family and/or group members to pose in ways that show how the director feels/experiences his/her family. Sculpting allows each person to "sculpt" the relationship in a way that allows others to visually "see" how the person experiences the relationship. It can help each family member recognize that though her or his individual perception and experience is clear, it is not the only way to view and experience the family. This insight can be powerfully experienced through sculpting.

Because sculpting is nonverbal, the information expressed in the family sculpture can often be more easily received by family members. It can be a way of describing relationships and feelings about the relationships that is powerful and effective.

Typically, sculpting occurs in two phases: the current family sculpture, and the "ideal" family sculpture. This exercise, therefore, begins with depicting the way the current family relationships are, with the second step depicting how the sculptor would like the relationships to be. Thus, this activity can result in the development of goals for the individual and family to help them move toward the desired family relationships.

OBJECTIVES

- To help families recognize the different perspectives of different people and acknowledge legitimacy of perspectives that are different than their own
- To help families develop respect for other family members' experiences
- To help parents recognize that their children have very different experiences than their own (and possibly are impacted more than parents know by substance abuse)
- To normalize strong feelings about family experiences
- To encourage hope in the possibility of change
- To identify at least one goal for families as a result of the sculpting exercise

MATERIALS

- No specific supplies, though members may want to use props, such as chairs to stand on

OPENING QUESTION SUGGESTIONS

- What is your favorite movie, and why?
- What is one thing you and your parents disagreed about when you were growing up?
- What is something you view differently than someone else in the room (or someone you work with or a friend or relative)?

METHOD

1. Explain that this is an exercise to understand how the individual "sees" and experiences his/her family. Talk about the idea that each individual in a family has his/her own unique experiences. Ask if anyone grew up with a sibling and realized later that they have very different memories and experiences. Ask if anyone is willing to share his or her experience. Normalize this experience (it happens in every family).

2. Explain that this exercise is a way to show what the relationships in your family are or were like. Let the group know that it involves asking people to stand as if they were in a sculpture. Let them know that the activity could involve being gently touched and "put" into a position, and that anyone is welcome to say they are not comfortable with being touched.

3. Ask for a group member to volunteer to sculpt his/her family. If there are no volunteers, the therapist can sculpt his/her own family as an example. If the group leader is sculpting as an example, use group members to represent various family members.

4. Give the volunteer the instruction to sculpt his/her family. If the family members are present, the group member can use them. If not present, ask the group member to choose people from the group to be the family members. This activity is most effective

when there are family members present, since each family member can sculpt his/her own experience.

5. Give the instruction: "You can move people in any position to show us what it was like to be in your family. You can tell them with words how to stand or sit, or, if they give you permission, you may move them gently with your hands. Show us what it is like for you to be in your family."

6. When they are done, ask the sculptor to tell the group about the family sculpture. Group leader can ask questions (how did that feel?) and make comments (you are looking away from your father)—but be careful not to make judgments (you must not be close with your father).

7. Ask each member of the sculpture what it feels like to be in the sculpture.

8. When finished, ask if there are any changes the sculptor would like to make in his/her family, and give him/her an opportunity to change the sculpture.

9. Then talk about how they could go about starting to make those changes in their lives.

10. If other family members are there, go through the same process with each of them. Make sure each family member has a chance to sculpt the family. If children are present, it is best to have the parents sculpt first.

GROUP DISCUSSION AND PROMPTS

- Normalizing family members having different experiences can be helpful. Group leaders can share their own experiences of how and when they realized this was the case in their families.
- Be sure that everyone obtains permission to touch each other before doing so; otherwise ask people to move into position with only verbal prompts.
- How did it feel to be moving people into position in your family sculpture?
- What did you learn from making this sculpture?
- Ask the people participating in the sculpture how it felt to be in that position.
- Make comments (not judgments) about the placement—you are furthest from your father, your sister is between you and your mother, your mother is looking away from you, and so on.
- If family members are present, ask if there were any surprises in the sculptures done by other family members.
- Make sure you give the opportunity to make changes in the sculpture, and ask what they could do in real life to make that change more likely.
- Thank participants for their willingness to participate.
- If family members are not present, ask the individual to take a guess at what the sculpture would look like from other family members' perspectives.
- If children are present, have the parents sculpt first, then the children.

AVOIDING PITFALLS

- Family sculpting often involves gently touching other people to move them into their poses. Group leaders and group members must be aware that this may not be comfortable for some people and must ask for permission before touching anyone.

- As with many experiential exercises, this activity can trigger strong emotions in participants. It is important to allow time to process those experiences. Moving slowly from one sculpture to the next and creating opportunities for processing are vital steps when facilitating this activity.
- As with other activities, group members should let the group know they are available to stay after group if anyone has been triggered.
- If families have a difficult time with the instruction to sculpt the family, group leader can give a more specific instruction to sculpt a particular event in the family (for example, "show us what dinner time looks like in your family" or "show us what happened when your parents fought when you were a child" or "show us what happens in your family when you are using").
- This activity is most effective when there are family members present, since each family member can sculpt his/her own experience. If they are not present, however, group members can choose other group members to "be" the family.

CULTURAL CONSIDERATIONS

- Group leaders must be aware of the possibility of offending people who are not comfortable with close proximity to others and/or of touching others. Anyone touching another person must do so only after first asking and obtaining permission. Respect must be given for those who are not comfortable with touching or being touched.
- Group leaders must keep in mind that many group members have a background that includes trauma and must be sensitive and respectful to those issues during this activity—again, particularly related to discomfort surrounding touching or being touched.

FOR GROUPS WITH YOUNG CHILDREN

- This activity is generally appropriate for groups with children.
- When sculpting families with children, offer the parents the opportunity to do the first sculpture (including what they would like to see different), and then the children.
- Ask the children (and parents) if there were any surprises in the sculptures.

4

My Family: Cycles and History

Activity Title: Genogram

Activity Mode: Worksheet/ Diagram/Drawing/Expressive Arts

RATIONALE

Many people with substance abuse problems come from families that had substance abuse problems. In fact, it is common to work with families that have generations of problems with substances. A genogram can be a helpful way to help group members and family members identify this (and other) cycles and/or patterns in their families. We think of "cycles" as ways of functioning that are passed from one generation to the next. For example, addiction often cycles through generations of families. "Patterns" are ways of functioning that occur repeatedly in the current generation. Recognizing patterns and cycles can come as a relief to individuals who feel guilt and confusion about where they are in their lives.

A genogram is a pictorial representation of at least three generations of a family (McGoldrick & Gerson, 1986). It can be extremely detailed and formal—though in this group activity, it is not meant to be so. This activity can help group members identify patterns and cycles as well as begin to identify ways of beginning to break patterns.

The activity can also be a way for family members to begin to tell their stories to each other and start to break some of the unhealthy patterns of keeping secrets the family may have developed. In our groups, we have seen this activity open a door to an exchange of stories and experiences in families that often had never happened before. Families sometimes go home planning to continue working on a more complete and formal genogram or other related activities, such as looking through old photos together.

OBJECTIVES

- To help families identify cycles (both helpful and not helpful) in their families
- To help families identify patterns (both helpful and not helpful) in their families
- To help families identify which cycles and patterns they want to continue and which they want to end

- To help families have the experience of telling their stories to other family members (or other group members if family members are not there)
- To encourage hope in the idea of "ending the cycle"
- To encourage hope in the possibility of changing old patterns
- To acknowledge family cycles and patterns related to addiction while avoiding blaming the family

MATERIALS

- Pencils/pens/colored pens, etc.
- Worksheet: What Is a Genogram?
- Large paper (to draw genogram on)

OPENING QUESTION SUGGESTIONS

- What is something you enjoy that your mother/father enjoyed (besides using)?
- When you were little, who in your family told you the best stories about your family?
- What is a similarity you see between your mother and grandmother (or father and grandfather)?

METHOD

The Genogram and Worksheet (about 30 minutes)

1. Ask the group if anyone is familiar with the idea of a genogram, and let them know it is a pictorial representation of our families that goes back at least three generations. It may be helpful to find and print a sample from the Internet.

2. Draw an example (see next page) on the board of a simple genogram, showing the various representations (male, female, age, married, divorced, etc.).

3. Ask the group to draw their own genogram. Family members may work together.

4. Ask each person to identify qualities, characteristics, or ways of living in each person (both positive and negative—drugs, alcohol, abusive, creative, musician, storyteller, divorced, early death, attorney, sexually abused, single mother, divorced, etc.). They may use different colors to code qualities if desired.

5. Hand out worksheet and have group members complete it.

GROUP DISCUSSION AND PROMPTS

- Gather the group into a circle.
- Ask the group what the experience was like for them.
- Ask the group what cycles they were able to identify. Ask if family members agreed on the cycles—were there any disagreements?

- Ask for examples of cycles (such as addiction, someone other than mother raising children, domestic violence) they want to end, and what they are doing to do so.
- Ask for examples of cycles they want their families to continue (importance of education, close connection in families).
- Ask for examples of patterns (identified in the same generation) they want to end and what they are doing about it.
- Ask for examples of patterns they want to continue in their family.
- Encourage storytelling. Encourage curiosity. Ask group members what it is like to draw their genogram. What did they learn? What do they want to find out more about?
- Ask for volunteers to share their genogram and the answers to their worksheet. Generally, there are some group members willing to share. If there is someone there with a family member who is willing, encourage them to talk about what the experience was like to work on the genogram together. Did they learn anything new about their family?
- Encourage each group members to share something about what their experience was during this activity.
- One way to close this group can be to ask the group members to share one glimpse of hope they have gained from this activity. Many members will talk about ending the cycle as a hopeful direction.

CULTURAL CONSIDERATIONS

- Group leaders should be aware of different cultural norms related to structure of families. In some cultures, for example, being unmarried with children is considered unacceptable, while in others it is more accepted. Being sensitive to any shame group members feel about the structure of their family is essential.
- There is some reading required with this activity, though if group leaders are aware of group members' limitations, this can be minimized by having the group leader review the symbols with the group.

FOR GROUPS WITH YOUNG CHILDREN

- This activity can be implemented successfully in groups with young children. Parents can be coached to tell family stories and praised for building a family culture with their children that encourages openness.
- Group leaders can model concern for content with children in the room. Many parents tend to believe that since their children have already been exposed to so much, they "already know everything." Group leaders can model appropriate boundary setting and help parents recognize that children should not "know" everything.

PITFALLS

- This project can seem overwhelming to some group members. It is important to emphasize that the genogram does not have to look any particular way; it is just a way of getting at information. Also, many people begin with a rough outline and then recopy the genogram when they have a more clear idea of the spacing and so forth.

- Group members sometimes tend to focus only on the negative patterns in their family and have a difficult time identifying some positive traits. Usually, with some encouragement and sensitive questioning (and examples from other group members), it is possible to help group members recognize some positive family traits.
- Group members can fall into the trap of blaming their family background for their situations. It is important to emphasize that although the cycle may be part of the "setup" for the addict, it is not the "fault."
- Parents often feel guilt about raising their own children in an addictive home. Group leaders should be sensitive to this and willing to talk about it in the group. Normalizing these feelings and being conscious of not shaming individuals is imperative.
- As with other family-related activities, the genogram can stir up strong emotions. Be aware of group members' responses to this activity, and let them know they can stay behind to process further if necessary.

GENOGRAM

1. Identify (on your genogram) qualities (both positive and negative, such as drugs, alcohol, abusive, creative, musician, storyteller, divorced, early death, attorney, sexually abused, etc.). Use different colors to code qualities if desired (this can be helpful in identifying repeating qualities).

2. Use the following genogram symbols to describe relationships in your family (close, conflictual, distant, cut-off, fused, living together/married/divorced).

3. Identify which characteristics, qualities, or ways of living are found more than once and seem to be passed down from generation to generation (cycles).

   _____

   _____

   _____

   _____

4. Which of these qualities do you want to maintain and pass on to future generations?

   _____

   _____

   _____

   _____

5. Which qualities would you like to stop passing on?

   _____

   _____

   _____

   _____

6. Are there any qualities in your generation that are patterns (even though not apparent in earlier generations)?

   _____

   _____

   _____

   _____

7. Are any of these patterns that you would like to end?

_____

_____

_____

_____

8. What are some ways to end the cycle (or the pattern) in your family?

_____

_____

_____

_____

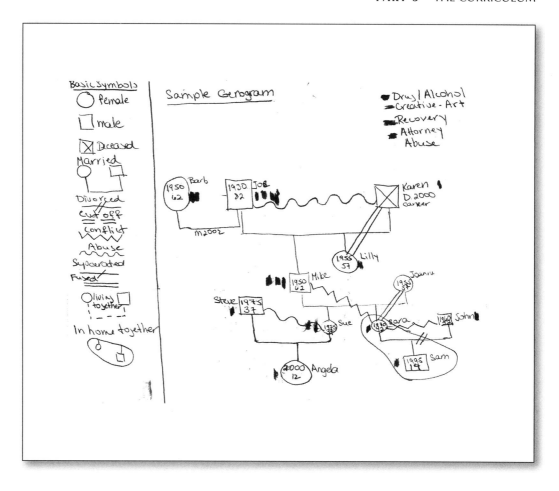

Curriculum Section II:
Family Identity

1

Family Identity I: Who We Think We Are

Activity Title: Family Coat of Arms

Activity Mode: Expressive Arts

RATIONALE

Family identity or family esprit de corps is a family's sense of who it is. It is composed of a family's experience and beliefs and often is passed down from generation to generation. "The Brown family sticks together" or "education is important in the Smith family" or "the Williams family doesn't hit" or "you don't cross the Jones family" are all examples of aspects of a family's identity. A strong sense of family identity can help provide guidance for individuals in the family and help create strong individual senses of identity.

Every family has an identity, though this is often not spoken of or acknowledged. Families with substance abuse often have a chaotic and undefined family identity and/or a family identity that is developed and maintained around the substance use. For example, many families hide their substance abuse from the outside world, creating a family identity that involves holding secrets, not talking about issues, and hiding substance use—as well as a family identity that includes everyone using substances in the family.

Because these rules and deep assumptions are often not talked about in families, children often do not recognize the possibility that there are other ways for families to be.

Part of the value of this activity is in creating a space and structure for clients and family members to begin thinking about their family's identity. Many unconscious assumptions family members have gained through their own family experiences are identified, expressed, and sometimes questioned and challenged, and new assumptions (building a new family identity) can be seen as a possibility.

In this activity, families identify pieces of their family identity that are helpful and that they want to maintain as well as areas they would like to modify. Family members are encouraged to talk about and process aspects of their family that often they have not talked about before.

Group members are then encouraged to identify a "family motto" that summarizes some important aspects of their family. This part of the exercise can be valuable as family members work together to create a shared vision of what their family means to them.

OBJECTIVES

- To introduce the concept of family identity or esprit de corps
- To help group members explore their own family identity
- To consider the possibility of creating a new family identity
- To point out that families tend to do things the same way their families of origin did, and it takes work to do things differently
- To encourage the group to look at what family identity they inherited from their family of origin
- To facilitate thinking about what pieces of their family identity they want to keep and what they want to change
- To inspire hope that modifying the identity of our families is possible

MATERIALS

- Family Coat of Arms sheet
- Family Coat of Arms questions sheet
- Crayons, markers, pens, pencils

OPENING QUESTION SUGGESTIONS

- What is something that is important to you in your life (besides recovery)?
- What is one thing you have changed about yourself in your life (besides stopping using)?
- What is something important you would like to tell your children (besides don't use drugs/alcohol)?
- What is something important your parents taught you?

METHOD

1. Tell the group that we are going to work on family identity. Ask them if they have any ideas what that means. Give information as appropriate (see rationale above).

2. Ask the group to provide some examples of their family's identity. (You can use some of the ideas in the above rationale to help them get started.)

3. Hand out the Coat of Arms sheets (2) and crayons, markers, and so forth. Let group members know they are going to create a "coat of arms" for their family. Tell them they can draw pictures or use words. Emphasize that there is "no right or wrong." Also let them know that if there are other areas important to their family that need to be on the coat of arms, they are welcome to add them. Have family members work in groups to create their family coat of arms together. If the family already has a coat of arms, instruct the family that they are going to develop a more current version of their coat of arms.

4. Once they have completed the coat of arms, ask the group to develop a "family motto"—a phrase that sums up something important they learned from their family (e.g., "nothing is stronger than blood" or "we support each other").

5. When they are done drawing, have them complete the questions on page 2 of the activity.

6. When people are done, gather them back into a circle for sharing. Ask what the process was like for them, and ask them to share what they wrote as they are comfortable.

MODIFICATIONS

- This activity can be done in two phases, with the first phase being a coat of arms from the family of origin and the second phase being a coat of arms from the current family. In this format, processing would include conversation about the differences between the two.
- A third phase could be added, with group members creating a coat of arms that includes a vision of how they would like their family to be. Discussion can include creating an action plan for getting there.

AVOIDING PITFALLS

- Group members may be hesitant to draw, so it may helpful in this activity to again emphasize that members can use any way of expressing their ideas. Having a supply of magazines that group members can use to cut and paste pictures from may be a helpful addition to the project. Using words rather than pictures is fine. Emphasize that this project is not about creating something that looks perfect but on creating something that expresses their experiences.
- Group members can be confused about "which family" to describe in their coat of arms (current family or family of origin). Although this activity is designed with the family of origin as the basis for the coat of arms, either family can be used. Particularly if children are present and working with the parents, it may be more helpful to work on the current family. In this case, parents can talk with the group about what they learned (brought with them) from their families of origin.
- As with all expressive activities, care must be taken by group leaders to assess for strong emotions that may lead to being triggered. Be sure to normalize this experience, ask if anyone is experiencing this, and offer to stay behind to meet individually with anyone feeling triggered.
- It is important to process the idea that there is no "perfect family" and that every family has negative and positive traits. Group leaders should take care not to portray group members' families as "unhealthy" or "dysfunctional." If someone does this with their own family, encourage them to see the dysfunction as a historical process they are working to modify.

CULTURAL CONSIDERATIONS

- This activity can be an excellent opportunity to bring forward the concept of different cultures having different ideas of what is important in families. Group leaders with members from a variety of cultures can use this opportunity to explore and acknowledge the values in different cultures. Modeling respect and interest in other cultures' values can encourage group members to show the same respect and interest.

FOR GROUPS WITH YOUNG CHILDREN

- This activity can be an excellent activity for families with young children. One way of working can be having parents work together with their children, with the parent working on a coat of arms from his or her family of origin and the children creating one from their current family. This method provides opportunities for parents to share some of their experience, as well as get information from their children about how the children experience their family.

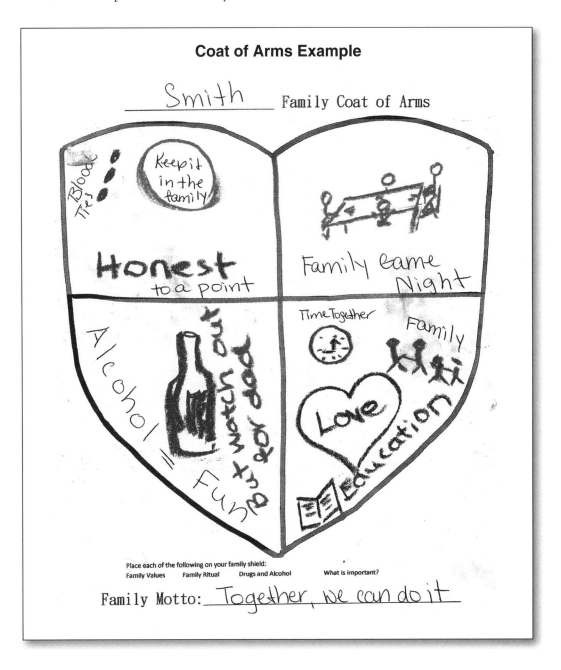

Coat of Arms Example

Family Coat of Arms

Place each of the following on your family shield:

Family Values   Family Ritual   Drugs and Alcohol   What is important?

Family Motto:_____

FAMILY COAT OF ARMS (CONT'D)

1. What do I want to take from my family of origin and carry into my current family?

2. What is something from my family of origin that I want to choose to let go of?

3. What do I want to replace that with?

2

Family Identity II: Structure

Activity Title: Family Routines and Rituals

Activity Mode: Psycho-educational (worksheet)

RATIONALE

Routines are day-to-day activities that are required for a family to function in a healthy way. Examples of routines are mealtimes, putting kids to bed at bedtime, homework time, and the like. Many families with substance abuse lack consistent routines and, as a result, feel chaotic and unstructured. Consistent routines allow for the family to accomplish basic tasks and help children feel safe.

Rituals are routines with symbolic meaning about family identity. These include holiday celebrations, spiritual/religious activities, family gatherings, and so on. Rituals are differentiated from routines in that rituals are linked to deeper meaning and senses of identity. As with routines, rituals are often not developed in families with substance abuse.

Both routines and rituals are important for families, and developing these can be a step toward a healthier family.

This activity serves to help group members take a step toward developing a more healthy family by identifying (sometimes creating) and strengthening family routines and rituals. Families often can identify unhealthy routines/rituals (such as a ritual of fighting on holidays or a routine of getting drunk every evening), and group leaders can help the families create a plan to move toward healthier ways of being. Many families we work with do not have healthy routines and rituals, and it is important to offer support and encouragement for creating them.

OBJECTIVES

- To introduce again the idea of family identity as an important foundation for a healthy family

- To differentiate between routines and rituals
- To help group members understand importance of routines and rituals in "healthy" families
- To help group members identify ways of creating healthy routines and rituals in their families
- To help group members clarify what they learned in their family and what routines and rituals they experienced in their family of origin
- To normalize a lack of positive routines and rituals in families with substance abuse
- To help families build a more positive identity
- To help families share their feelings and experiences
- To acknowledge and normalize strong feelings that can arise as group members explore their family history

MATERIALS

- Family Routines and Rituals worksheet
- Pens, pencils

OPENING QUESTION SUGGESTIONS

- How were meals prepared and eaten when you were growing up? What was/is your family's favorite celebration?
- What was your routine at bedtime when you were growing up?
- What is something important that your family usually does together?
- When you were young, what was your routine for getting ready for school in the morning?

METHOD

1. Ask the group what "family identity" means to them.

2. Introduce ideas of routines in families and rituals in families. Ask the group to explain the difference (see rationale above for explanation). Ask for examples of routines and rituals to help group members clarify the differences. Use the examples of group (routines include circling up, introductions, processing time, closing quote, etc.; rituals include acknowledging a group member graduating).

3. Present routines and rituals as important in developing a healthy family identity.

4. Hand out the Family Routines and Rituals worksheet, and give the group some time to complete the worksheet.

5. Ask the group to share their responses.

6. Point out how the experiences of routines and rituals in their own families helped them develop their own family identity.

7. Continue with the idea that there may be some routines/rituals they want to keep for their own family and some they may want to let go. Not all routines/rituals are positive/healthy (for example, the routine of children going to their room when Dad starts drinking or the ritual of Christmas being a negative time once Mom starts drinking).

8. Ask group members to think of one new routine they would like to begin in their family, and one ritual.

AVOIDING PITFALLS

- There may be group members that are not able to identify positive routines and rituals in their families. This experience could trigger strong emotions such as guilt or sadness. Acknowledging and offering support for this experience is essential. This is an opportunity to allow group members to plan new routines and rituals for their family. Point out that creating a routine can be as simple as having a meal together every day (or once a week). Each routine will give the family more structure. Helping group members to see making these changes as possibilities (not too overwhelming) is essential.
- Group members sometimes will tend to blame their families and become very angry. It can be helpful to encourage group members to use this anger as motivation to make changes in the future—to do things differently.

CULTURAL CONSIDERATIONS

- Expressing interest and curiosity about rituals from different cultures can help group members be willing to share their experiences. Modeling this interest and curiosity can be helpful for the group.

FOR GROUPS WITH YOUNG CHILDREN

- This activity can be adapted for groups with young children. Children may be able to identify routines that parents are not aware of (for example, children may tell each other stories after going to bed, or children may comfort each other when parents are using). Helping parents create an atmosphere in which children are not afraid to share these experiences is essential. Group leaders can always offer to meet with parents afterward to process some of these issues.
- This can be a good opportunity for parents to learn more about their children's experiences.
- Children do not always want more routine (such as an earlier, regular bed time). Parents should be encouraged to set boundaries that are healthy in the long term for their children and that may not always be what the child wants in the moment.
- Parents in recovery often feel guilty about not being there for their children in the past. Group leaders can normalize these feelings while still emphasizing the importance of boundary setting (including stable routines).

FAMILY ROUTINES AND FAMILY RITUALS WORKSHEET

Family Routines:

Daily:

Weekly:

Other:

One routine I would like to change/add:

Family Rituals Worksheet:

When I was growing up:

My current family rituals:

One ritual I would like to change/add:

3

Family Identity III: Who We Feel We Are

Activity Title: "My Family" Poem

Activity Mode: Creative Writing (worksheet)

RATIONALE

The "My Family" poem activity allows group members to work individually or as families to create a poem that describes various attributes of their family. It is a fill-in-the-blank activity that allows group members to be creative and encourages them once again to think "out of the box" about their family.

Many group members come to group with a background of trauma and many childhood memories that are painful. The writing that results from this activity can often express some of those painful memories. On the other hand, group members often create a piece that is funny and lighthearted. Our experience is that reading this poem to the group results in positive feedback that encourages and supports the group members' expressions of their own experiences.

This activity can serve to remind group members of some experiences they have not thought about in some time—and, as with other creative activities, it can trigger strong emotions. This can be a jumping-off point to help group members frame their experiences in ways that can be helpful and again consider what they want their current family to look like.

OBJECTIVES

- To continue to help the group explore the idea of family identity
- To help the group explore their own family identities
- To help individuals and families work on developing their own sense of family identity
- To help individuals clarify what they learned in their family
- To help group members share their feelings and experiences

MATERIALS

- "My Family" sheet
- Pens, pencils

OPENING QUESTION SUGGESTIONS

- If your family were a group of animals, what would they be and why?
- How many brothers and sisters do you have? Where are they now? Are you in touch with them?
- Who is someone in your family that you admire? Why?

METHOD

1. Hand out "My Family" worksheet and pens.

2. Ask the group to complete the worksheet. Families may work together or separately.

3. Ask the group to not take this exercise too seriously. They can write down the first thing that comes to their minds. If there is one that is difficult, they can leave it blank. There is no right and wrong. Be creative.

4. Ask the group for volunteers to share.

GROUP DISCUSSION AND PROMPTS

1. Ask the group what their experiences were as they wrote the poem and if any surprising feelings, memories, and so forth came up for them.

2. Ask what it was like for the families that worked together on the project.

3. Ask what the group learned about their family identity.

4. Ask what their thoughts are about changing one's family identity—is it easy or hard to change? Is it possible? How?

AVOIDING PITFALLS

- This is an activity that requires reading and writing. Group leaders must be aware if there are clients in the group that do not have those skills and help them process the activity in a way that does not shame them. Some group members are willing to sit with a group leader and work with him or her. Many group members feel badly about not reading, and group leaders must be adept at helping and supporting in ways that are not offensive.
- Group members, particularly those that have not reached a point where they are willing to fully participate in group, can sometimes become very negative about their families during this exercise. In most groups, this can be balanced by encouraging input from group members that are strong members of the group. If group members become disruptive, group leader can ask the group to do the activity in silence.

CULTURAL CONSIDERATIONS

- This activity can be difficult for individuals who do not speak English fluently or who speak English as a second language. Group leaders can emphasize there is no right or wrong, and part of the creativity of this exercise is the unusual ideas that often arise. Group leaders expressing interest and encouragement for this creativity can be helpful modeling for the group.

FOR GROUPS WITH YOUNG CHILDREN

- This can be an excellent exercise to use when there are children participating in the group. Children can bypass the intellect much more easily, often, than adults can, and the result can be wonderfully creative and insightful poems.
- One way of facilitating this exercise is to have the parents create one poem and the children create their own, then have each share.
- Another option is to have parents and children create a poem together. This can give helpful insight into how the parents and children negotiate different opinions and desires and work together.

"MY FAMILY" POEM

By _____

My family is _____

My family feels _____

My family pretends _____

My family wonders _____

My family sees _____

My family hears _____

My family wants _____

My family worries _____

My family cries _____

My family understands _____

My family does not understand _____

My family says _____

My family dreams _____

My family tries _____

My family hopes _____

My family believes _____

My family is _____

4

How I See My Family

Activity Title: Circle Drawing

Activity Mode: Expressive Arts

RATIONALE

It is common that though family members are raised and live in the same family, they each have very different experiences of how their families are. In families with substance abuse, this can be even more pronounced. This activity can serve to help each individual clarify how he/she experiences the various relationships in the family. When family members are present, the exercise can help with communication that bridges those diverse experiences.

The Circle Drawing (Edwards, 2003) is a simple activity that can help group members become more aware of how they experience their families and how other family members experience the family. It can also help individuals develop a realization of how they would like to experience their families (goals)—what they would like to be different. This can lead to concrete steps they can take to move toward that goal.

OBJECTIVES

- To help individuals gain an understanding of how they experience their families
- To help individuals acknowledge that each family member has her or his own unique experience of the family
- To help individuals recognize the impact of relationships in their family they may not have been aware of
- To help individuals and families think about how they would like to experience their families
- To help individuals identify what they can do to move more toward how they would like to be in their families and how they would like their relationships to be
- To help individuals share information with family members and to increase openness to others' experiences

MATERIALS

- Large sheets of paper
- Pencils or pens

OPENING QUESTION SUGGESTIONS

- Who would you want to be closer to in your family?
- Can you think of an event in your life that you remember differently than another family member does?
- If something wonderful happened in your life, who would you tell first?

METHOD

1. Instruct the group to draw a large circle on their paper.

2. When the circles have been drawn, ask the group to "place your family on the paper." Tell them they can place them close together or far apart, large or small, and to use circles to represent each person (do not draw heads, arms, legs, etc.).

3. Write names (or relationships such as "dad," "mom") in the circles.

4. Family members should not see each other's drawings as they draw.

5. When finished, gather the group together again, and ask who would like to share. Each group member (one at a time) holds his or her drawing up and describes what he/she drew (who is in his/her family). Group leaders can use suggestions from below (Group Discussion and Prompts) to comment on the circle drawings, which will generally inspire discussion from other group members.

6. When discussion is over, ask each person to turn their paper over and draw their family the way they would like them to be.

7. Again, make comments and encourage sharing. Ask, "What is different?"

8. Ask each participant to identify one action he or she can take to move the family closer to how he/she would like it to be.

GROUP DISCUSSION AND PROMPTS

- Make observations about each drawing, ask questions, and prompt discussion. Do not make interpretations (such as "you must be closer with your mother than your father"); just make comments (such as, "I notice that you are closest to your father and your sister closest to your mother" or "your father is much bigger than you are").
- Ask, "Is there anything here that surprises you?"
- If family members are present, ask what the differences are between different family members' drawings. Is that surprising?
- If family members are present, also ask what the similarities are in the drawings. Are those surprising?
- Ask the group to notice who is in the drawing (often people draw pets or other important relationships).
- Did anyone draw circles outside the large circle? What does this mean?
- What did you learn from this activity?

AVOIDING PITFALLS

- The most common pitfall in this activity is for group leaders to give too much instruction. It is important to give the instruction as directed above, and when the group members ask for more instructions (which they will), give only the same instructions. Part of the effectiveness of this activity is that each individual creates the circle drawing in her or his own way. Further instruction inhibits this process. One way of responding to repeated questions is to remind them that there is no wrong way to complete this activity.

- When describing the activity, give instructions one step at a time. Do not let group members know they will be drawing their family as they would like it to be until they have finished drawing and processing the drawings of their current family.
- A way of thinking of the circles in this activity is that the circles "are" the people in the families (rather than a representation of the people). This way of thinking can help group leaders formulate their responses in ways that are very concrete. For example, once a group member has told you that the large circle she has drawn is her father, you refer to the large circle as "your father." The group leaders might point out that "your father is very close to your sister, but not so close to you"—then pause to wait for a response. This method of responding helps keep group members present and concrete with this activity rather than becoming abstract.

CULTURAL CONSIDERATIONS

- Different cultures define "family" in different ways, and these differences may be expressed in the circle drawings. This can be an opportunity for group leaders to model curiosity, interest, and respect for cultural differences, creating a group culture that encourages the expression of differences.
- This is an excellent activity for groups with members that do not read or write, as it requires neither skill. Rather than writing names, family members can be represented by different colors or shapes in the drawing.

FOR GROUPS WITH YOUNG CHILDREN

- This activity is appropriate for groups with young children. Each individual in the family can be encouraged to draw his or her own circle drawing and then share them with each other. This can give good information to parents about their children's experiences and possibly a direction for treatment for families that are willing to work on these issues.

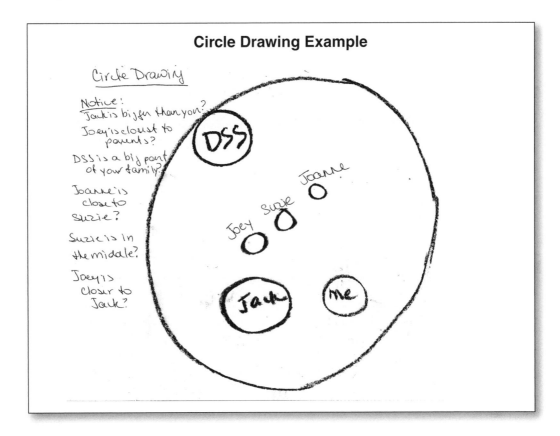

Curriculum Section III:
Sober Fun

1

Family Fun

*Activity: Group/
Family Game Night*

Activity Mode: Game

RATIONALE

One symptom of substance abuse is that people often lose the ability to have fun without the use of substances. In many families, times of celebration always involve heavy substance use. In fact, times of celebration can often be scary and uncomfortable times for children and nonusing family members.

Game night is generally a group favorite. We often facilitate this activity at a time when the group feels unstable (often when the group is "new"), as game night can serve to strengthen the cohesion of the group. In addition, this activity provides another opportunity for people to have fun without using drugs or alcohol—a new experience for many group members. It also serves to model a family game night, which we encourage families to incorporate as a ritual in their own family structure.

In general, we use games that can be modified to be played with two teams. Many purchased games (such as Pictionary®) can be easily modified in this way. Often, group members will want to suggest games; encouraging this can be a way to increase group members' participation in and feelings of ownership of the group.

OBJECTIVES

- To learn or relearn how to have fun without drugs and alcohol
- To build cohesion in the group
- To have fun together
- To emphasize the possibility and the importance of creating fun family times
- To introduce idea of "family game night" ritual for families

SUPPLIES

- A game (see below) and whatever is necessary to play the game

OPENING QUESTION SUGGESTION

- What is one fun thing you did with your family when you were growing up?
- **Caution:** This question may bring up bad memories and difficult feelings. Frame the question with an understanding that many of our childhoods were difficult and painful, and we have lots of bad memories—but tonight we are going to talk about one good thing, even a small one.
- What was your favorite game growing up?

METHOD

1. Let the group know that we are going to play a game together. Catch Phrase™ is one that works well with a small or large group. Charades™ or Pictionary™ work well with two teams. (See below for more detailed instruction on how to structure the game.) Do not use game boards, just alternate drawing/acting and guessing. If a group member leads the group, that person can bring any game he or she wants. The key is to make it simple and fast moving—adapt the rules of any game so it is played with two teams.

2. PLAY THE GAME! If required for the game, ask for a volunteer to keep time and to keep score.

3. Circle up again and open a discussion (see below for prompts).

    Note: We often invite the group to have a "covered dish" night on Game Night.

GROUP DISCUSSION AND PROMPTS

- What was it like to play the game?
- How does playing the game relate to families? (Don't know how to play at first, but figure out as it goes along, families have fun together, taking turns, team felt like a family, etc.)
- Did anyone have a family game night growing up? What do you think of that idea? Is anyone going to try it with her or his current family or friends?
- What are other ways to have fun without drugs or alcohol?

HINTS FOR GAMES

Catch Phrase™: Follow game instructions. If it is a large group (10 or more people), keeping the teams clear can be challenging. One trick is to arrange chairs in a circle with every second chair placed in a bit, so the group can see which team is playing. Make sure that people are close enough that the game can be passed from person to person easily.

Pictionary™: Use the Pictionary cards and timer. Divide the group into two teams. One person draws from one team (using a whiteboard at the front of the room). Drawer picks a card and draws as many pictures from that card (both sides) as possible, before timer runs out.

Drawer's team guesses and gets one point for each correct guess. When timer runs out, switch teams.

Charades: Use a purchased charades game, or make up ideas as part of the group activity. Split the group into two teams. Teams take turns having a member act out charade and get one point for each correct answer.

AVOIDING PITFALLS

- This is an activity that can be fun and exciting—it is not unusual for group members to sometimes contribute inappropriate comments or forget to monitor their language. This can be lessened by a reminder from the group leader before the games begin and as the emotions increase in the room. This is particularly important when children are included in the activity.
- There may be people in the group for whom it is painful to remember childhood experiences. Group leaders need to be sensitive to this and remind group members that they can stay after to group if anyone is upset or triggered.

CULTURAL CONSIDERATIONS

- This can be another opportunity to model respect and interest in other cultures, as group members may bring experiences from their childhood that are foreign to others. Expressing genuine curiosity and interest is essential.

FOR GROUPS WITH YOUNG CHILDREN

- This is an excellent activity for groups with young children. It is important, however, to remind the group that young children are present, and it is necessary to monitor their language. As excitement increases as the game progresses, this is sometimes difficult for people to remember. Group leaders should emphasize the importance of children not being exposed to harsh language.

Playing Together

2

Activity Title: Drumming/Music

Activity Mode: Expressive Arts/Music

This activity encourages "sober fun." Recognizing that having sober fun is something many group members have not done for many years, we recognize the importance of building this experience into the family component of substance abuse treatment. The instructions below are very general and may be adjusted as needed to create a fun music event. Be prepared that group members will feel self-conscious (as can group leaders)! This activity does not require a strong musical orientation by group leaders, though recruiting group members that have some experience with drumming can be both inspirational and helpful.

OBJECTIVES

- To continue to learn or relearn how to have fun without drugs and alcohol
- To build cohesion in the group
- To have fun together
- To emphasize the importance of creating fun family times
- To introduce idea of playing music together as a ritual for families

SUPPLIES

- African drums, percussion instruments (rattles, shakers, etc.), empty 5-gallon buckets to use as drums, any other instruments and noisemakers
- Printed words to several well-known songs (accessible on the Internet). You can ask the group for suggestions as you plan this activity.
- Invite the group to bring musical instruments and choose the songs.

OPENING QUESTION SUGGESTIONS

- When you were growing up, what place did music have in your family?
- What does music mean to you now?
- What is your favorite song?
- Who is your favorite artist?

METHOD

1. Acknowledge that playing music in public can be difficult, and ask if this activity makes anyone nervous? Talk about the importance of being willing to take risks (being in recovery is taking a risk). If it is true, let the group know that you are also nervous!

Part I—Drumming

1. Have the group sit in a circle, with each person choosing an instrument. It is important to let the group know that this activity is not about being musical or playing music well; it is about trying something different, listening, and having fun together.

2. The group leaders should divide the group down the middle, into two groups. Have one group keep a beat with their drums (bang on the beat, 1, 2, 3, 4...). Have the second group bang twice with each beat (1 & 2 & 3 & 4 &...). Let the group that this is the beginning of drumming. Encourage them to keep the rhythm, and as they feel moved to branch out and drum in different rhythms, feel free to do so.

3. Let the group know they can always come back to this simple beat (1, 2, 3, 4) if they lose the rhythm.

4. It is helpful if one person leads the drumming. Encourage the group to listen, and modify what they are doing as the group changes

5. If there is a group member that is familiar with drumming, it is helpful to recruit the member to start the drumming with a simple, regular beat. Encourage group members to join in as they feel comfortable.

6. Generally, the drumming will build and find a natural end and each round of drumming will stop without direction from the leader. If that is not the case, the group leader can use his/her hands to motion to quiet and slow the beat.

7. After each round, ask for people's experiences.

Part II—Singing

1. Hand out the preprinted lyrics to two or three well-known songs. This is another time when it is helpful to have a group member step up to lead the songs.

2. Emphasize this is not about singing well, it is about trying something new together and having fun.

Part III—Performing

1. If there are some group members that brought instruments and are interested in performing, allow time for that.

GROUP DISCUSSION AND PROMPTS

- Present drumming and music making as something we can do together to help us connect with each other. Ask the group to share some of their experiences with music growing up or recently.
- At the end, circle up and ask those present to share their experiences. It can be helpful to talk about the difficulty of shyness, and express appreciation for the group being willing to participate in this activity.
- This can be an intimidating group to lead. Sharing the group leader's fears and anxiety can be helpful in the group as modeling a willingness to take risks and try something new.
- Ask the group how many are uncomfortable performing music, and acknowledge their willingness to try it, even though they are uncomfortable.
- Ask if anyone played music with their families growing up.
- Ask if anyone would consider having a family music night in their home.

AVOIDING PITFALLS

- Often, people will say playing and/or experiencing music is linked to drugs and alcohol for them. It is important to acknowledge this issue, offer support and encouragement, and point out that in this group, we are learning to do many things without using that we used to do while using. Also emphasize importance of talking about triggers, and let the group know they will be able to do so at the end of group as well.
- Many group members will also have used in the past to help them face uncomfortable, difficult situations. Group leaders should acknowledge this and acknowledge the group's willingness to try something new, even if it is uncomfortable.
- Group leaders are often uncomfortable leading this activity. This is a good opportunity for them to model dealing with those uncomfortable feelings—talking about group leader's nervousness can be a helpful normalizing experience for group members.
- Group members may respond to their discomfort by not participating. It is helpful to have a number of musical noisemakers (spoons, rattles, etc.) that will allow a way for members to participate yet not feel too much in the spotlight.
- Again, it is important to normalize feelings of being triggered during processing time and for group leaders to offer to stay after group if anyone feels the need to do so.

CULTURAL CONSIDERATIONS

- This can be a wonderful opportunity for group members to share music from their cultures. Group leaders could consider asking group members to bring examples of music from their culture. (If group members bring music, a guideline to not bring music that glorifies drug and alcohol use should be given.)

FOR GROUPS WITH YOUNG CHILDREN

- This can be an excellent activity for groups with young children. Children are usually eager to participate in the musical activities and can model the joy of creating music for the adults in the room. Parents can have the experience of creating quality fun time with their children.
- If there are older children, the group leader may be able to recruit them to help lead the music, particularly drumming.

Curriculum Section IV:
Toward Health

1

Healthy Helping

Activity Title: Healthy Helping Worksheet

Activity Mode: Worksheet

RATIONALE

Many families with substance abuse struggle with knowing how to help the addicted family member. They want to know what is going "too far" and how to set appropriate boundaries. The commonly used term *codependent* is often, in our experience, felt to be insulting by family members, who, as they point out, love the addicted family member and are very genuinely trying to help but just don't know how.

This worksheet provides a noncondemning approach to identifying what we call "unhealthy helping"—areas in which family members try to help the substance abuser but do so in ways that actually work to maintain the addiction. The activity aids family members in identifying some traits of unhealthy helping (such as feeling guilty when taking care of themselves, lying to protect the addicted family member, etc.). Families are also encouraged to identify actions they can take to toward more "healthy helping."

The activity also provides an opportunity for the addicted person and the family members to talk about these issues, which often have not been spoken of previously.

During this activity, families often hear the stories of other families and recognize that they are not the only ones struggling with these difficult and painful issues. This can be an experience of great relief for many families.

OBJECTIVES

- To introduce concept of healthy helping versus unhealthy helping and encourage discussion and sharing experiences. Many families will be aware of the term *codependence* and have that as a negative concept. Frame codependence as "unhealthy helping"
- To normalize "unhealthy helping" in families, particularly families with addiction. Point out that everyone has some of these traits, and it is a matter of maintaining balance
- To acknowledge the desire to help family members, the difficulty in identifying when helping has gone too far (become unhealthy), and the difficulty in stopping these behaviors

- To help families identify steps they can take to move toward healthy helping
- To create an environment in which families can get support as they move toward setting limits and creating boundaries

MATERIALS

- Healthy Helping worksheet
- Pens, pencils

OPENING QUESTION SUGGESTIONS

- What was the last thing you did to help someone else?
- Who has helped you in your life?
- What is something you can do to "give back"?

METHOD

1. Introduce the topic: healthy helping and unhealthy helping. Ask the group for their ideas of what this means. What is healthy helping? What is unhealthy helping? Ask group for examples from their own experiences.

2. Hand out Healthy Helping worksheet and pens. Give some time for everyone to complete the worksheet. Ask people to do this part of the exercise individually.

3. Ask group members to share what questions stood out to each person and whether they identified any unhealthy helping patterns in their family.

4. Ask each person to share one thing they identified that they will stop doing and one thing they will start doing, (or that they will ask their families to stop or start doing) to move toward healthy helping.

GROUP DISCUSSION AND PROMPTS

- If family members are not present, group members can complete this exercise using examples of healthy/unhealthy helping they receive from their families. They can take the step of talking about this topic with their families at home and making requests of their families to help them move toward healthy helping. Group leaders can suggest that families come to a family therapy session to further address this topic.
- Ask family members to identify times in their lives they wanted to help or thought they were helping but really were helping in a way that is unhealthy.
- Ask group members to identify times their family tried to help them but ended up helping in a way that is unhealthy.
- Normalize this! Families need to hear that they are not "bad" for unhealthily helping their addicted family member. This is normal and happens with all families.
- Emphasize that these issues are excellent issues to be pursued in therapy sessions.

AVOIDING PITFALLS

- Many people will be familiar with the terms *codependence* and *enabling*. These terms are sometimes experienced as blaming the family for continuing the addiction, and this topic can bring up strong feelings in family members. It is important for group leaders to acknowledge that *enabling* is a result of trying to help someone you love. Family members' intentions are good, even though enabling can be hurtful in the long term.
- This is also a good time to talk about the benefits of Al-Anon for family members. Group leaders should have a schedule of Al-Anon groups in their area to give to family members.

CULTURAL MODIFICATIONS

- Because different cultures have different definitions of what is healthy helping, it is essential for group leaders to be sensitive to what is a cultural issue and what is a healthy/unhealthy helping issue. The group leader can explore with the family the culture of the family and help the family decide what is healthy or not.

FOR GROUPS WITH YOUNG CHILDREN

- In general, this is not an activity for groups with younger children. However, this activity could be used in groups with older (teenage) children and could be useful in helping the children identify what healthy and unhealthy helping are and how they appear in their family. Children raised in a family with alcoholism or drug addiction are more likely to develop those patterns themselves, and recognition of healthy/unhealthy helping patterns can be helpful in intervening in that process.

HEALTHY HELPING WORKSHEET

	Yes	No
1. I have a hard time saying no, so I tend to do things even after I have said I won't.	___	___
2. I often feel guilty when I say no, even if I know that saying no is reasonable.	___	___
3. I do things for others that they could do.	___	___
4. I feel guilty when I do something for myself.	___	___
5. I do more for others than they do for me.	___	___
6. I put others' needs before mine.	___	___
7. I lie to protect people in my life.	___	___
8. I do things to protect others from real-life consequences.	___	___
9. I help others keep secrets, or I keep secrets.	___	___
10. I seem to want my loved one to get better more than he/she does.	___	___

Are there other ways I am unhealthily helping?

_____

_____

_____

_____

What changes, if any, do I need to make in ways that I help? Are there things I need to stop doing? Are there other things I need to start doing?

Things I need to STOP: Things I need to START:

_____   _____

_____   _____

_____   _____

_____   _____

_____   _____

_____   _____

Doing It Differently

Activity: Poem: "New Hands"
by Carol Lynn Pearson; 9
Dots Exercise; My Family's New Story

Activity Mode: Expressive Arts
(writing); Puzzle Solving

RATIONALE

We all tend to live what we know. Children growing up in families with abuse often grow up thinking abusive relationships are normal. Children growing up in families with substance abuse learn that chaos is normal. These patterns are often unconsciously transmitted from one generation to the next.

Breaking these patterns involves, among other things, expanding one's perception of possibilities—in other words, recognizing that something else is possible. Often our clients exclaim, "I didn't know there was any other way" or "that was normal for me" or "I thought that was how everyone lived."

One step in the process of moving toward recognizing and opening to new possibilities is recognizing the current (and past) patterns. Once patterns are identified, clients can choose whether they want to maintain the pattern in their family or change it. We speak of this as "ending the cycle," a powerful concept for many parents who want their children to have positive lives but do not know how to make that happen.

This exercise introduces the idea of thinking "outside the box," using a puzzle that group members can work on individually or in teams. To solve the puzzle, group members need to think outside of the boundaries the puzzle seems to be in and think outside the box.

We then use a poem to help introduce the idea of changing patterns in families, and that highlights the big impact these changes can have on our children.

Finally, we facilitate an expressive arts (writing) activity that helps clients tell a story about our lives with an ending they would like.

OBJECTIVE

- To help group members understand the impact of family experiences on our lives
- To normalize repeating patterns and offer support for parents who feel they have hurt their children

- To encourage thinking about the impact of childhood experiences
- To encourage parents to want their children to tell a different story
- To help group members consciously envision another ending to the story
- To stimulate thinking about how parents can help their children tell a different story
- To encourage hope about the possibility of breaking the cycle

MATERIALS

- Paper and pencils/pens
- "New Hands" poem handout
- "Nine Dots" exercise
- "New Story" handout

OPENING QUESTION SUGGESTIONS

- What is something (besides using) you had to stop or start doing, and was it hard to do?
- Have you ever had an "aha" moment?
- What is something you did recently (besides staying sober) that you feel good about?
- What is one thing you love about your family (either current family or family of origin)?

METHOD

1. Hand out Nine Dots exercise and give time for group to complete.

   a. Say, "Connect all the dots using only four straight lines, without lifting your pencil from the paper."

   b. Hint: You have to think "outside the box."

   c. It is fine for people to work together on this project. Give them about 5 minutes, then ask if anyone solved the puzzle.

   d. Ask, "What was that like?" Ask the people who were successful how they figured it out.

   e. Point out that to solve the puzzle, you have to think out of the box. Relate this to going out of your comfort zone (as in recovery) and doing things in a way that is completely different.

   f. Ask group members for other examples of times when they have had to think outside the box (or when they have had to change).

2. Read poem (or ask for a volunteer to read): "New Hands" by Carol Lynn Pearson

   a. Ask for feedback and thoughts from group, and encourage discussion about impact of substance abuse and violence on children.

   b. Ask what group members felt as they read/heard the poem. Ask what their children might feel about the poem.

   c. Talk about patterns in families, difficulty identifying and changing patterns, and normalize possible pain and/or guilt in families, particularly those with substance abuse.

3. Facilitate "Once upon a time" exercise.

   a. Describe the exercise and give time for the group to write their stories (e.g., Once upon a time…and the story changed when…and now…). As with many expressive arts exercises, the less verbal instruction, the better. Let the group know that as long as they participate, there is no way to do it wrong.

   b. Allow about 10 minutes (more or less, depending on how quickly people are done).

   c. Ask for volunteers to read their stories.

   d. As each person shares, give him or her some time to talk about what the process of completing the story was like.

GROUP DISCUSSION AND PROMPTS

- Encourage discussion of patterns that are handed down in families. Substance abuse is one example, but group members will be able to identify many more. Emphasize that not all patterns are negative—positive patterns are transmitted to the next generation also. It can be helpful to have the group identify some of those positive patterns. Examples can be a tradition of camping together, going fishing, playing games, and so forth.
- It is important that when group members share their stories, the group leader also asks them what the process of writing the story was like.
- It should not be required for group members to share their stories, but it can be helpful to go around the circle and have each member say something about her or his experience with these exercises.
- Tables will probably be necessary to write on, but when the stories are completed, the group should come back into a circle to share.
- Once the three parts of this session are completed, the group leaders can ask the group why they think the three sections (Nine Dots, poem, and story) are grouped together.

AVOIDING PITFALLS

- Group leaders need to be aware if any group members are not able to read or write. If so, group leaders can sit with those persons and help them with the story. Group leaders must be aware that this can be embarrassing to those group members, and they must handle this situation with care. Having a conversation outside of group with any group members who are not able to read/write can be helpful in eliciting ideas from the group members of how the group leader can help.
- A writing activity can be intimidating to group members (as can all expressive arts activities). Group leader should emphasize that this is not about being a "good" writer, and there is no way they can do it wrong (other than not participating).
- Strong emotions can result from this exercise, as well as from the poem. Group leaders should emphasize that they are willing to stay after group if strong emotions arise for group members.
- Group leaders should normalize feelings of guilt and regret when we talk about parenting in group. Most people with substance abuse have guilt about how they have

parented their children. We cannot change the past, but we can do it differently right now. Acknowledge group members' courage in facing these issues and thinking about how they will do it differently.

- When asking for group members to share, it can be helpful to go around the circle and say something about what this activity was like for them. Group leader can emphasize that though they would like to hear what group members have written, no one needs to share more than they are comfortable with.

MODIFICATIONS

- The three activities in this session are grouped together and meant to be implemented during the same session. This is not, however, by any means essential, and any of the activities can be implemented on its own.

CULTURAL CONSIDERATIONS

- Because this exercise requires writing, group leaders must be aware (as mentioned above) if there are any group members that do not write and be conscious about addressing this issue in a way that is respectful and discrete.
- Talking about patterns in families can be a good way to learn more from group members about their cultures' traditions. For example, many families in rural Appalachia sing together. Showing interest and respect for such cultural traditions can help group members be more comfortable about sharing.

FOR GROUPS WITH YOUNG CHILDREN

- The Nine Dots exercise can be a fun puzzle for parents and children to work together on.
- Group leaders should consider the age of the children and appropriateness of the poem. Group leaders can feel free to choose another poem that talks about children learning from their parents and then teaching what they learn to their children. Time can be spent on this topic, and group leaders can encourage children to talk about what they learn from their parents. This discussion will be different, of course, with children of different ages.
- The story is a simple enough and familiar enough concept with most children that even young children are able to develop a story. The story does not have to be complex or long. Most children are familiar with "once upon a time" and can identify something in their own life they can write a story about.
- If the children are not able to write, group leaders can sit with them and help them write their story.
- It can be a powerful statement for parents to read their own stories aloud so that their children can hear. This can be a way for parents to let their children know that it is okay for them to talk about these issues. At the same time, group members must be sure that the content of the stories is appropriate for children to hear.

New Hands

By Carol Lynn Pearson

Celia got drinking from her mother
And hitting from her father
And yelling from both
Like she got pizza crusts for breakfast.

And she took it all in
And digested it and became it
Because you are what you eat.

And her parents
Ate from the table of their parents
Who ate from the table of theirs
Back and back and back
And Celia was stuck.

But cells die
And every seven years we are new.

Celia's new heart and new hands
Set the table and stir the pot
And serve better stuff than she ever got.

Source: Carol Lynn Pearson, *Women I Have Known and Been.* Gold Leaf Press (1993).

THE NINE DOT EXERCISE

Instructions: Connect all nine dots using only four straight lines, without lifting the pencil from the paper

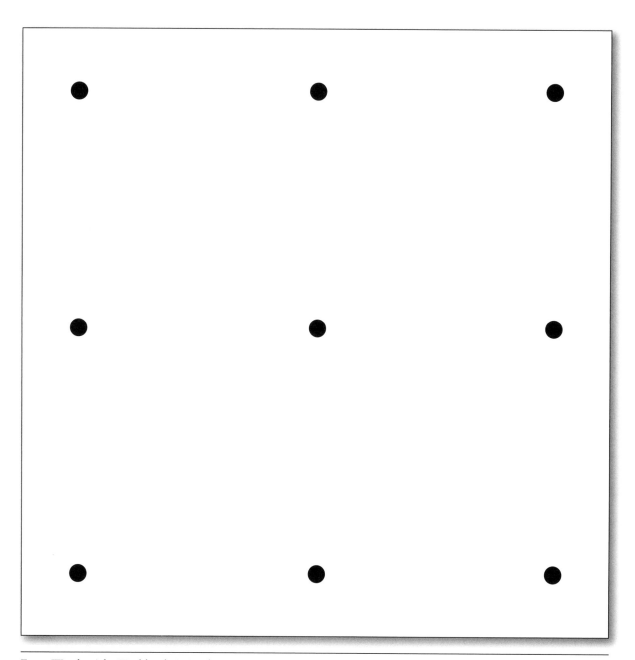

From Watzlawick, Weakland, & Fisch (1974).

The Nine Dot Exercise

Instructions: Connect all 9 dots using only 4 straight lines, without lifting the
pencil from the paper

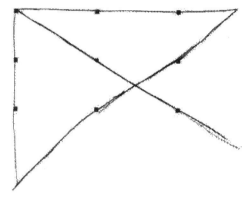

MY FAMILY'S NEW STORY

ONCE UPON A TIME,

AND THE STORY CHANGED WHEN,

AND NOW,

My Family's New Story Example

My Family's New Story

Once upon a time, my children were happy. They played and laughed, and didn't know what was coming.

The story changed when my husband died, and I left my children in another way. I drank and used and did anything to not feel pain. It didn't work.

And now, I am back. Sometime I feel my broken heart, but I know that is ok. I am back with my children. I am sober + I plan to stay that way. This story has a happy ending!

Relationships I: Healthy or Unhealthy?

Activity Title: Healthy Relationship/ Unhealthy Relationship

Activity Mode: Psycho-educational (worksheet)

RATIONALE

Many people in our groups have no experience of a healthy relationship. Most readily acknowledge that this leaves them unprepared to choose anything other than what they know—an unhealthy relationship. Many of our clients have moved from one unhealthy relationship to another. Substance abuse tends to aggravate this pattern, as drug and alcohol usage and sexual acting out are often linked. The result is an intensification of the inability to create a relationship that is positive.

This activity involves completing two worksheets. The first helps group members identify healthy and unhealthy ways of being in a relationship. The second helps identify patterns and a direction toward changing those patterns.

The goal is to help group members begin to evaluate their current and past relationships and begin to develop some critical thinking about the relationships they choose. They are encouraged to identify patterns in the types of relationships they choose and consider whether there are patterns they would like to change. In addition, the ability to choose a healthy relationship is presented as something we all need to learn. Hope is encouraged, and the fact that many of us do not grow up with good models of healthy relationships is normalized.

OBJECTIVES

- To help group members begin to conceptualize healthy relationships
- To normalize the challenge in creating healthy relationships

- To point out that many of us did not have models of healthy relationship as we grew up, so it makes sense that we do not choose healthy relationships
- To help group members assess health of their current relationship, if they are in one, or past relationships if they are not
- To help group members identify patterns in the relationships they choose
- To help group members identify directions they want to move in toward improving the health of their relationships
- To inspire hopefulness about the possibility of creating healthy relationships

MATERIALS

- Pens, pencils
- Worksheets:
  - Healthy Relationship/Unhealthy Relationship
  - What I Experienced Impacts My Relationships Today

OPENING QUESTION SUGGESTIONS

- Do you know someone who is in a relationship you see as healthy? Who?
- What is something you think is important in a good relationship?

METHOD

1. Introduce topic of healthy versus unhealthy relationships. Ask what the group thinks are qualities of healthy relationships (it is helpful to make a list on the board if possible).

2. Ask the group to share experiences of who they had in their lives as they grew that modeled healthy relationships. Normalize the fact that many people do not have these healthy role models, so it makes sense that we struggle with knowing what a healthy relationship is and how to create one.

3. Hand out the Healthy Relationship/Unhealthy Relationship worksheet and give the group time to complete it. Facilitate discussion about what they learned about their relationships. (If people are not in a relationship currently, ask them to use their past relationships.)

4. Ask about patterns group members see in the relationships they choose. Give the group time to process what they recognize, as well as ways in which they would like to change those patterns. Normalize people developing patterns based on their experiences in life as both children and adults. Facilitate a discussion of the difficulty of recognizing and changing those patterns.

5. Hand out What I Experienced Impacts My Relationships Today worksheet, review questions, and give some time for group members to consider responses. Allow time for sharing, particularly responses to the last question—what they can do to move in the direction of having the relationships they want.

GROUP DISCUSSION AND PROMPTS

- It can be useful to normalize the fact that many group members have not experienced many healthy relationships. Talking about this can serve to inspire hope that future relationships can be healthier—that we can learn to do it differently.
- It can be helpful for group leaders to acknowledge that many people in the group probably had difficult experiences growing up and have not had a chance to learn to do things differently. Acknowledging their willingness to learn and change can also serve to inspire hope.
- It can be useful to ask each group member to identify one thing he or she will do differently as a result of this exercise.

AVOIDING PITFALLS

- Many group members may not have experience with healthy relationships in their lives. It is not uncommon for members to say they don't know what that means or they wouldn't recognize a healthy relationship. Normalizing this experience and instilling some hope that this does not mean they cannot have a healthy relationship is important.
- Group leaders should be cautious that if family members are in the room and there is any domestic violence in the relationship; individuals may be at risk if they are honest about their relationship. It is essential that group leaders respect this risk and not push people to share more than they are comfortable with. If there is any concern about domestic violence, group leaders should meet individually with clients (perhaps at another time) and assess for safety. Safety must always be the priority.
- It can be very helpful to end this activity on a hopeful note (one thing they can do).

MODIFICATIONS

- This activity can be implemented either individually, or group members can work together.
- The checklist can be presented as something that parents can use to help their children learn about healthy/unhealthy relationships.

CULTURAL CONSIDERATIONS

- Different cultures have different expectations about relationships. Normalizing cultural differences can be helpful.
- Group leaders should use genuine curiosity and interest to encourage group members to talk about cultural differences in relationships.

FOR GROUPS WITH YOUNG CHILDREN

- The checklist can certainly be reviewed with children and is a good way to help children begin to think about healthy/unhealthy relationships. Group leaders need to be aware that many children in the room may have witnessed violence and/or abuse, and these issues may come up during this activity.
- The worksheet is written at an adult level, so group leaders and parents may want to rework the worksheet to be more appropriate for children, paying attention to friendships the children are developing.

MY RELATIONSHIP: HEALTHY OR UNHEALTHY?

Although there are no perfect relationships, some are more healthy than others. Taking a look at the following may give you more information about the relationship you are in currently. If you are not in a relationship currently, answer these questions according to previous relationships.

If you are in a healthy relationship, you:

- ❐ Take care of yourself and have good self-esteem (independent of your relationship)
- ❐ Maintain and respect each other's individuality
- ❐ Maintain relationships with friends and family
- ❐ Have activities apart from one another
- ❐ Are able to express yourselves to one another without fear of consequences
- ❐ Are able to feel secure and comfortable
- ❐ Allow and encourage other relationships
- ❐ Take interest in one another's activities
- ❐ Do not worry about violence in the relationship
- ❐ Trust each other and be honest with each other
- ❐ Have the option of privacy
- ❐ Have respect for sexual boundaries
- ❐ Are honest about sexual activity if it is a sexual relationship
- ❐ Accept influence (allow give and take)

Your relationship may be unhealthy if you:

- ❐ Neglect yourself or your partner
- ❐ Feel pressure to change who you are
- ❐ Feel pressure to quit activities you usually/used to enjoy
- ❐ Pressure the other person into agreeing with you or changing who they are
- ❐ Notice one of you has to justify your actions (e.g., where you go)
- ❐ Are forced to share everything
- ❐ You or your partner refuse to use safe sex methods
- ❐ Notice arguments are not settled fairly
- ❐ Experiencing yelling or physical violence during an argument
- ❐ Attempt to control or manipulate each other
- ❐ Notice your partner attempts to control how your dress and/or criticizes your behavior
- ❐ Do not spend time with one another
- ❐ Have no common friends or have a lack of respect for each other's friends and family
- ❐ Notice an unequal control of resources (food, money, home, car, etc.)
- ❐ Experience a lack of fairness and equality

Source: Adapted from http://depts.washington.edu/hhpccweb/content/clinics/family-health/healthy-vs-unhealthy-relationships

WHAT I EXPERIENCED IMPACTS MY RELATIONSHIPS TODAY

Growing up, the relationships around me were: _____

_____

_____

_____

From that, I learned: _____

_____

_____

_____

I tend to choose relationships that: _____

_____

_____

_____

One thing I would like to change about that is: _____

_____

_____

_____

Even if I did not grow up with healthy relationship models, I can still create healthy relationships in my life. In order to do this, I can:

_____

_____

_____

_____

Relationships II: Choosing a Relationship

4

Activity Title: Choosing a Relationship Worksheet

Activity Mode: Psycho-educational (worksheet)

RATIONALE

Many of our clients have a history of being in unhealthy relationships. In fact, it is not uncommon for group members to acknowledge that they have never been in a relationship they believed was healthy or even knew anyone who was in a healthy relationship. Many of our clients grew up in homes with violence, substance abuse, and/or other negative relationship patterns. Many have only had harmful relationships modeled for them and have no concept of what a healthy relationship would look like (let alone how to create one).

Helping clients acknowledge and normalize this experience, thus normalizing their own pattern of being in unhealthy relationships, can feel like a relief to many clients. The idea of being able to change this pattern can encourage hope.

This activity serves to help group members recognize patterns of falling into relationships rather than using critical thinking to assess the person and make a conscious decision whether to be in the relationship. Group members are encouraged to identify traits that they want in a relationship, as well as traits they want to avoid, and to use that information when considering becoming involved in a relationship.

For those that are in relationships, this list of traits can serve as a direction for improving the relationship.

OBJECTIVES

- To normalize the experience of not recognizing a healthy relationship and falling into unhealthy relationships
- To help group members conceptualize healthy relationships
- To encourage the idea that it is important to choose a relationship rather than fall into one

- To help group members develop a clear idea of traits they want in a partner, as well as traits that are not acceptable
- For those that are currently in relationships, this list of traits can present a direction for improving the relationship
- To encourage hope in the idea that a healthy relationship is possible

MATERIALS

- Pens, pencils
- Worksheet: Choosing a Relationship

OPENING QUESTION SUGGESTIONS

- What are some qualities you think are important in a relationship?
- Who do you know that you think has a healthy relationship? Why?
- Who do you know that you have a healthy relationship with (romantic or platonic)?

METHOD

1. Pass out Choosing a Relationship worksheet and introduce topic. Ask for experiences of choosing a relationship rather than falling into a relationship. Ask group members if their parents or anyone else talked to them about how to choose a good relationship, and normalize our difficulty in doing so given that very few of us were encouraged in this idea growing up.

2. Give group time to respond to questions on the worksheet and then review them.

3. This exercise works well in small groups, as individuals sometimes experience difficulty thinking of character traits they want, and it can be helpful to have people work together.

4. For those that are in relationships currently, present the worksheet as a way to identify important qualities and growth areas for their relationship. Emphasize this is not meant to be negative about current relationships (all relationships can improve; it does not mean they are not good).

GROUP DISCUSSION AND PROMPTS

- Ask the group to share stories of people they know whose relationship seemed healthy and positive.
- Ask the group if it is difficult for them to be alone (not in a relationship). How does this impact the relationships they choose?
- Ask if there are particular relationship patterns they tend to choose (such as choosing abusive relationships, choosing partners with substance abuse or legal problems, choosing abandoning or critical partners). Ask if this is something they would like to change.
- Encourage the group to write this list of qualities on paper and keep it with them. How would that help them avoid falling into a relationship that resembles their old patterns?

- Ask the group if they find these ideas helpful and if they feel hopeful about changing their old patterns.
- Ask parents what they want to model for their children.

AVOIDING PITFALLS

- Many group members will be in relationships currently. It is important that they do not feel that group leaders are saying those relationships are not healthy. Group leaders should present this as an opportunity to strengthen current relationships.
- Group members sometimes feel hopeless as they recognize their own patterns and choices and the impact these have on their families and children. It is important to acknowledge these feelings and help people move toward some feeling of hope in the possibility of change.

MODIFICATIONS

- This exercise can be modified for use with people currently in relationships by focusing on things to change to help improve their current relationships. As they develop ideas of what they want in a relationship, for example, help them identify steps they can take to create that in their relationship.
- Couples can work together to create a vision of the relationship they want to create together.

CULTURAL CONSIDERATIONS

- Being cognizant of and respectful of the different cultural norms for relationships can be important in this exercise. It can be helpful to acknowledge that everyone has different ideas of what a healthy relationship is, and many of these ideas are culturally developed. Though there are some basic concepts we can all agree on (such as no physical violence), it can be very interesting for group members to hear what individuals experience in different cultures. Group leaders can model curiosity and interest in hearing group members share their cultural experiences, thus helping create a supportive, nonjudgmental group environment.

FOR GROUPS WITH YOUNG CHILDREN

- Some of the issues that come up with this topic (such as parents' experiences with violent relationships) can be inappropriate for groups with children. However, the idea of choosing relationships—and helping children identify what they want in a relationship and what a healthy relationship looks like so they can begin to choose their friends with these ideas in mind—can be important steps in helping them conceptualize and begin to practice choosing relationships in a healthy way.
- Much of this worksheet can be completed by children in terms of friendships (qualities they want in a friendship, etc.).

CHOOSING A RELATIONSHIP

What does it mean to CHOOSE a relationship rather than fall into one? Which do I usually do?

_____

_____

_____

_____

What are five qualities I want in a partner?

1. _____

2. _____

3. _____

4. _____

5. _____

What are four qualities that are NOT acceptable to me in a partner?

1. _____

2. _____

3. _____

4. _____

What are five things I want to be able to truthfully say about my relationship?

1. _____

2. _____

3. _____

4. _____

5. _____

Is now a good time for me to be in a relationship? Why or why not?

_____

_____

_____

_____

How will I know I am ready to be in a relationship (if not in one currently)?

_____

_____

_____

_____

What are things I want to do differently in my next relationship (or current relationship if already in one)?

_____

_____

_____

_____

5

Relationships III: Stages in Relationships

Activity Title: Stages in Relationships Worksheet

Activity Mode: Psycho-educational (worksheet)

RATIONALE

Many of the clients in our groups cycle through relationships quickly and often. It is not unusual for someone to leave a hurtful relationship, meet someone else, and be living with this new person in a very short period of time. For many of our clients, this pattern is "normal"—it is what they saw as they were growing up, it is what they see in the relationships around them, and it is how their relationships have always been.

The purpose of this activity is to present another view of relationships, a developmental perspective that shows that each stage of a relationship is based on the previous stage. Using this model, we can help clients identify their own patterns in relationships and also recognize that it could be helpful in the long term to allow their relationships to progress through the stages rather than jump from the beginning to end stage. These are new ideas to many of our clients that can help them view their relationships differently. They can also help clients develop hope for and a vision of a future that could include a healthy relationship that has progressed systematically through relationship stages.

This worksheet reviews one model of stages in relationships (Campbell, 1983). We emphasize that developing relationships is a process that takes time, and often people jump into relationships so quickly that the natural development does not have a chance to occur (for example, meeting someone and moving in with her or him 2 weeks later—long before there has been a chance to move beyond the "infatuation" phase of the relationship). Group members are encouraged to consider their own patterns of entering relationships and how this impacts the development of their relationships. They are also encouraged to identify patterns they would like to change, as well as one thing they can do to move in that direction.

OBJECTIVES

- To give information about stages in relationships and help group members identify stages in their own relationships
- To normalize changes in relationships based on the developmental stage of the relationship
- To help group members identify their individual patterns of entering and being in relationships and what they want to change
- To help group members begin to conceptualize healthy relationships
- To help group members identify concrete steps they can take to move toward healthier relationships in their lives
- To develop hope in the possibility of healthy relationships, even though many people have not had that experience

MATERIALS

- Pens, pencils
- Worksheet: Stages in Relationships

OPENING QUESTION SUGGESTIONS

- Who do you know that has what you would call a healthy relationship? What is one thing you respect about that relationship?
- Where did you learn your own patterns of being in a relationship?
- What is an important quality you want to have in a relationship?

METHOD

1. Ask group if they have experienced that relationships go through stages. Use group experiences to review Stages in Relationships worksheet.

2. Ask a volunteer to read the points under each stage. Facilitate a discussion of the group's experience with each relationship stage. Give time at the end of each section for them to write down what their experience is and what they would like to change.

3. Once all stages have been reviewed and discussed, give the group some time to think and write about their own relationship patterns and what they would like to do to move toward more healthy patterns. Ask the group members to share their thoughts.

GROUP DISCUSSION AND PROMPTS

- Let the group know that this could be new information for many people and that we are not taught this information in school. Ask who is surprised by this information.

- Ask who has experienced jumping quickly from an early stage (such as infatuation) to a late stage very quickly.
- It can be helpful to have the group talk about where they learned about relationships. Often, group members will share that they have never learned about healthy relationships.
- Ask the group how using drugs and alcohol has impacted their choices and experiences surrounding relationships, including decision making.
- Ask how loneliness impacts these same choices, experiences, and decisions.
- What about sex? How does having sex impact the progress of a relationship?
- Ask each person to share one thing she or he wants to try to do differently as a result of this exercise.

AVOIDING PITFALLS

- Many group members (often most group members) will not have had experiences with healthy relationships, either seeing healthy relationships growing up or having their own healthy relationships. Thus, there may be group members who have no experience of many of these relationship stages. It is important to normalize this experience and emphasize the importance of understanding this developmental process in order to move toward healthy relationships.
- There will likely be group members that are currently in a relationship. For these people, it can be helpful to ask them what stages they feel they have completed successfully and what stages they need to work on. For example, if they are living together and fighting, group leaders could normalize this (power-struggle stage, end-of-honeymoon stage) and introduce the possibility of continuing to work on these issues. Letting clients know that this is not unusual and can be helpful.

MODIFICATIONS

- The groups we run are made up of a combination of people, some of whom are in a relationship, others who are not. This activity is presented for such a group. It can be altered based on the makeup of the group. If the group in entirely made of couples, group leaders can focus on helping individuals identify stages they have not completed and how they can go back to more thoroughly complete that process. For example, a couple who married after knowing each other 2 months will likely be experiencing arguing or boredom. Normalizing these experiences and helping the couple identify ways to address these issues in constructive ways can be helpful. Often, helping individuals see that these experiences make sense in this developmental framework can feel like a great relief.

FOR GROUPS WITH YOUNG CHILDREN

- In our culture, we tend to not teach children about healthy relationships. From this perspective, this is an excellent activity for children to be involved in so they can begin to grasp the notion that relationships progress, change, and grow. Also, the importance of choosing relationships consciously is an important concept for children to grasp.

Third, the concept of taking steps to improve relationships can be an important one for children to understand.

- Children can use their relationships with their friends (what stages do your relationships with your friends go through? How do you learn to trust your friends? What happens when you fight with your friends? When did you know that _____ was going to be a good friend? What was different about your relationship with _____ a year ago? What would you like to see improve in your friendships with _____?) to begin to consider stages in their friendship relationships.

- In general, group leaders need to help parents set boundaries surrounding what they share with their children about their relationship. For example, children can know (and it would probably help them to know) that their parents are working on not fighting as much. Details of parents' fights should not be discussed with the children. Families we work with generally have structural looseness, and they need help in identifying what should be talked about in front of the children and what should be kept between the parents.

- One option would be to have this group be with parents only; then, as homework, have the parents take the information home to share with their children.

RELATIONSHIP STAGES

Phase 1: The Honeymoon-Romance Stage (Everything is Wonderful!)

- Strong attraction
- Romantic, passionate
- Everything is perfect
- Can do no wrong
- Love and belonging
- Short-lived (2 months–2 years)
- Focus on similarities, differences denied

Describe a couple you know with this style of relationship: _____

_____

_____

_____

My experience with the honeymoon stage is: _____

_____

_____

_____

One thing I want to change about this is: _____

_____

_____

_____

Phase 2: Power Struggle (Disillusionment)

- Differences emphasized
- Things that first attracted you drive you crazy
- Need to accept that differences are okay and normal
- Prerequisite to relating to each other as whole people
- Need to learn to fight fairly with both winning
- Need to declare one's own individuality and separateness

Source: The above was compiled from *The Couple's Journey*, by Dr. Susan M Campbell.

Describe a couple you know with this style of relationship: _____

_____

_____

_____

My experience with the power struggle stage is: _____

_____

_____

_____

One thing I want to change about this is: _____

_____

_____

_____

Phase 3: Stability

- Aware of each other's personal world, and the difference is okay
- Resting time; war is over
- Risk of moving apart
- Sometimes feeling of boredom, of not being connected
- You have history together, and it can be used to advantage
- Caution not to throw away relationship; can learn mutual respect or go back to Phase 2

Describe a couple you know with this style of relationship: _____

_____

_____

_____

My experience with the stability stage is: _____

_____

_____

_____

One thing I want to change about this is: _____

_____

_____

_____

Phase 4: Commitment

- Real readiness for marriage (though most marry earlier)
- Many couples don't make it to this stage
- Wide awake, making choices based on individual differences and things you have in common
- Balance of love, belonging, fun, power, and freedom
- Don't need each other; choose to be with each other
- Need to stay in stability phase until both ready for commitment

Describe a couple you know with this style of relationship: _____

_____

_____

_____

My experience with the commitment stage is: _____

_____

_____

_____

One thing I want to change about this is: _____

_____

_____

_____

Phase 5: Cocreation

- A team in the world
- World may include children, projects, church, business, etc.
- Danger is overinvolvement in the outside world, neglecting relationship

Describe a couple you know with this style of relationship: _____

_____

_____

_____

My experience with the cocreation stage is: _____

_____

_____

_____

One thing I want to change about this is: _____

_____

_____

_____

In relationships, I know that one pattern I have that I would like to change is:

_____

_____

_____

_____

One thing I can do to change this pattern is:

_____

_____

_____

_____

Note: Stages are not linear; rather, they are circular or spiral. You are in one stage or another at any given time, with bits of others. Knowledge of the stages helps you move through them.

The above was compiled from *The Couple's Journey* by Dr. Susan M Campbell.

6

Private or Secret?

Activity Title: Family Bag

Activity Mode: Expressive Arts

RATIONALE

Families with substance abuse often have family secrets—secrets that are not talked about and that individuals collude (verbally or nonverbally) with other family members to keep secret. An example of this would be Dad's drinking not being talked about openly, and no one in the family bringing up the issue. Mom may call in to work for Dad in the morning, saying he is sick. Kids may make sure they are not around after dinner, since that is when Dad starts drinking heavily. The family silently works to avoid bringing up the issue and protect the family secret.

All families have issues that are private. Private issues are not talked about for healthy reasons (such as not talking about details of parents' sexual relationship), and not talking about them does not hinder the health of the family. Secrets, as in the above example of the family keeping Dad's drinking secret, are kept quiet to maintain an unhealthy system.

This is an expressive arts activity that encourages families to identify and think about family secrets and differentiate between privacy and secrecy. Group members are encouraged to think about these differences and create a collage (using magazine pictures, drawing, writing) to express what their family shows on the outside and what their family keeps hidden. Pictures are placed on the outside of a paper bag to show what the family displays to the outside world. Inside the bag, group members place parts of their family that are kept hidden.

As part of this process, families may think and talk for the first time about these ways their families function. This can be a profound experience. When group members have family members present, this activity can trigger conversations about family issues and dynamics that have not previously been spoken of.

OBJECTIVES

- To point out that though all families have secrets, families with addiction often have many large secrets
- To distinguish between private issues and secrets
- To offer the opportunity to talk about family secrets
- To normalize difficulties in families (no "perfect" family)

- To help group members identify patterns in their families they want to change and a plan to move toward changing those patterns
- To encourage people to think about what they may want to change in terms of secrets in their families and how to do that
- To encourage people to think about how their children may think about family secrets

MATERIALS

- Brown paper lunch bags
- Magazines (to cut from)
- Glue
- Scissors
- Markers, crayons, pencils

OPENING QUESTION SUGGESTIONS

- When you were growing up, what was something about your family that you were proud of?
- When you were growing up, what was something about your family that embarrassed you?

METHOD

1. Introduce idea of family secrets (all families have secrets, but often there are more in families that have addiction). Talk about the family face that is shown to the outside world and the family face that is only shown to family members.
2. Ask the group to explain the difference between private issues and secrets. Private issues are not talked about for healthy reasons (such as not talking about details of parents' sexual relationship), and not talking about them does not hinder the health of the family. Secrets, as in the above example of the family keeping Dad's drinking secret, are kept quiet to maintain an unhealthy system.
3. Ask group for examples of their own family secrets that they are willing to share (if no one volunteers, give an example yourself).
4. Have the group complete the "before the project" questions on the worksheet.
5. Describe this activity as an opportunity to think about the secrets in your own family and the different qualities and traits that are shown to the outside world versus what is shown inside the family.
6. Describe the bag as the family, and instruct group members to glue pictures/words to the front of the bag that represent how the outside world sees the family. They can also draw/write on the bag.
7. Pictures/words that represent what is *not* shown to the outside world go inside the bag.
8. Make the suggestion that families work together to create their bag, but this is not a requirement.
9. Let the group know they will be asked to share, but they need only share what they are comfortable with (some or all of their project); it is okay to pass, but it may be

helpful to have members who do not want to share their project talk about what the experience was like for them.

10. Allow about 20 to 25 minutes for making the project.
11. Have the group complete the "after the project" questions on the worksheet.
12. Then gather the group back to the circle and facilitate sharing the family bags.
13. If time allows, as people share, ask them what it was like to do this project—did they think about anything they hadn't thought about before?

GROUP DISCUSSION AND PROMPTS

- This exercise can be facilitated with group members using either their family of origin or their current family as the family they work with. Either way, it can be helpful to ask what some of the differences are between the family of origin and the current family.
- If people are hesitant to share what is inside the bag, it can be helpful to ask them what it was like to think about those things. Was it difficult? Did it stir up any emotions?
- The question of what people's experiences were during this exercise is always helpful. Memories and emotions can be triggered, and normalizing this possibility is important.
- Ask group members if they have thought about the difference between privacy and secrecy, particularly in their family. What are they teaching to their children? How can they open the door for their children to talk about difficult subjects?

AVOIDING PITFALLS

- At some point, it is good to point out that this project could bring up strong feelings, and encourage people to talk about that (either in the group or after) if that happens. Normalize "imperfect" families (no such thing as a perfect family) and explain that one of the goals of this exercise is to encourage them to think about what they want to maintain in their family, as well as things they might want to change.
- Group members who are struggling with maturity can sometimes attempt to get attention from the group by sharing drug and/or sex-related projects. It can be a challenge for the group and group leader to view these attention-seeking behaviors as diagnostic information and respond appropriately—which could mean ignoring the behaviors or redirecting them toward more appropriate participation.

MODIFICATIONS

- This activity can also be facilitated as a "bag of self" exercise, using the outside of the bag to show what we as individuals show the outside world and the inside of the bag to show what we do not show the outside world.
- The family working together on one bag often creates a different bag than when each individual works on his or her own bag. It can be an interesting modification to have individuals in a family work on their own, then come together to create one project that includes everyone's experience. This can also be a good way to acknowledge that everyone in the family has a unique experience.

CULTURAL CONSIDERATIONS

- Group leaders should be sensitive to cultural differences in terms of what is deemed appropriate to share and what is not. Asking and expressing curiosity and interest about cultural rules about privacy can be helpful and can initiate important conversations about how we learn what we practice.

FOR GROUPS WITH YOUNG CHILDREN

- This exercise can be an excellent one for families to work on together. Group leaders can help parents work with their children to create a family bag. Part of this will mean coaching parents to encourage children to share experiences and feelings that they may not have shared before. It can also be a time for parents to share some of their experiences growing up with their children.
- It may be difficult for children to express issues that have been secrets in the family. Parents can talk to the children about changes they are making in their lives and discuss that talking about experiences (even difficult and painful experiences) is okay.

FAMILY SECRETS

Before the project:

What is the difference between something that is secret and something that is private?

_____

_____

_____

_____

What is an example of a family secret from your family (when you were growing up)?

_____

_____

_____

_____

After the project:

What was it like to work on this project? Did you think about anything you hadn't thought about before?

_____

_____

_____

_____

Are there any secrets in your family that you feel it may be helpful to talk about with your family? If so, what?

_____

_____

_____

_____

Do you find that your comfort level with secrets is similar to what you grew up with, or is it different? In what ways?

_____

_____

_____

_____

Families in Recovery I

Activity Title: What's True and What's Not? Family Goals and Relapse Issues

Activity Mode: Psycho-educational (handouts/worksheet)

7

RATIONALE

Many family members we work with are relieved when we frame addiction as a "family disease" that impacts not only the individual substance abuser but also everyone in the family. We refer to "early recovery" as a stage not only for the substance abuser but for the family as well. We identify unhealthy patterns family members have become entrenched in as the family learned to function around an addiction, as well as goals for moving toward changing those patterns. We talk about common experiences, "myths" family members often believe, and goals for the family during various stages of recovery.

This group's activity consists of two psycho-educational handouts. The first (Families in Recovery: What's True and What's Not?) describes some common experiences and beliefs of families in early recovery. Individuals are encouraged to differentiate between beliefs they have that are true and those that are not true. (For example, many family members believe they can either cause the addict to relapse or stop it from happening. This is identified as a myth.)

The second handout (Family Goals and Relapse Issues) outlines goals for families in three different stages of recovery and helps family members identify specific actions that would be helpful for them to take as their using family member continues in recovery. It also presents relapse risks for family members and helps family members identify coping skills that could help reduce the risk of a family relapse.

It can be very helpful for family members to learn that many of the experiences they have while their loved one starts on a path to recovery are common. Many people in our groups have expressed relief to learn that they are not the "only ones."

OBJECTIVES

- To give families information about families in recovery
- To help families distinguish between the addict's recovery and the family's recovery

- To help families identify some myths and some truths about the addict's recovery and their role
- To help individuals develop goals to progress in the family's recovery
- To open the door for families to talk openly about these experiences (such as being hypervigilant)
- To normalize these experiences and help families feel they are not alone

MATERIALS

- Handouts:
  - Families in Recovery: What's True and What's Not?
  - Family Goals and Relapse Issues
- Pens, pencils

OPENING QUESTION SUGGESTIONS

- How do you know when you're doing too much for someone else?
- What is something that has changed dramatically since you (or your family member) stopped using?
- Who is someone you would like to say thank you to and haven't? What for?

METHOD

Families in Recovery handout, What's True and What's Not?

1. Distribute handout.

2. Ask for a volunteer to read the "true" statements.

3. Encourage discussion, particularly between family members. Ask if anyone in the group has experienced this (chances are most have). Be sure you get input from both the addict group member and family members.

4. Point out that these are very common experiences. Most family members can relay stories of their own hypervigilance.

5. Ask the group how they cope with this experience. Frame these as coping skills.

6. Ask a volunteer to read the "myth" statements.

7. Continue the discussion. Ask the group if there are any other myths that should be on the list.

8. Give some time for them to complete the worksheet (how these ideas show up in their family, how they can change their thinking, and what they can do differently).

9. Ask for volunteers to share what they wrote.

Family Goals and Relapse Issues

1. Distribute handout.

2. Present recovery as a family issue, not just an issue for the addict. Say that when there is addiction in a family, the family survives in ways that often are not healthy. For example, when Dad drinks, the children may go to their rooms and stay there. There may be unspoken rules in the family that when Dad is "sick," everyone must be very quiet to avoid further problems. When Dad stops drinking, it is sometimes hard for other family members to know how to act! Point out that if one person in a family changes, the whole family must change. This exercise is about helping family members identify goals for their family's recovery and a plan for reaching those goals.

3. Give time for group members to complete the worksheet, including the rating scale.

4. Ask group members to identify actions they can take to achieve the goals. Ask them to identify things they can do to help them move up to the next number on the scale. (If you are at a 3, what could you do to get to a 4?)

5. Point out the relapse issues and encourage the group to talk about these issues, as well as ways to lessen the risk of these issues, and other issues that could be risky for their family in particular.

GROUP DISCUSSION AND PROMPTS

- Much of the value of this exercise (as with many) can come from the discussion between family members. One of the most common experiences we see is that family members often feel a great relief when they recognize that they are not alone in their experiences—in fact, their experiences are shared by many. This is a good time to mention Al-Anon and encourage family members to attend an Al-Anon meeting.
- In the section asking "how this shows in my family," it is helpful to ask group members to be very specific (When Johnny comes home at night, I want to find ways to get close to him to smell his breath). Once a specific incident is reported, it is easier to identify changes to make in both thinking and doing.
- Group leaders can frame many of the goals of later recovery as something to think about for the future (depending on where the addict is in recovery). For groups that work with clients in early recovery, much of the focus can be on the goals for early recovery.
- It can be helpful to ask the group how they feel about framing the family as well as the addict in recovery.
- Families can work together on completing both these worksheets.

AVOIDING PITFALLS

- Some family members may take offense at the idea of the family as well as the addict being in recovery. It is important to not give families the idea that we are saying the family members are being blamed for the addiction. Talking about systems and how one part of the system impacts all the other parts of the system can be helpful.

- Most of the groups we run are for people in early recovery. It may not be helpful to identify action steps to reach the goals for the later stages of recovery (particularly stage 3, 20+ weeks) at this point in time.
- Group leaders must be aware that the topic of making decisions about relationships (leave or trust) can be a volatile one. Normalizing this issue can be helpful, as well as presenting that decision as an ongoing process rather than something that must be decided at a certain point in time.
- This topic works best when there is a high percentage of family members in the group.
- Group leaders must be aware if there are group members who cannot read and write. In such a case, group leaders can speak with those individuals outside of group and ask how they can be helpful. In our group, group leaders often sit and work with individuals, particularly those that we recognize need help with reading and writing.

MODIFICATIONS

- For groups that struggle with low numbers of family members, these worksheets could be given to individuals as homework sheets to take home, work on with their families, and bring back.
- These worksheets could also be used with individual families in family therapy sessions.

CULTURAL CONSIDERATIONS

- Group leaders need to be sensitive to cultural differences in the structure of families. What could be considered "enabling" in one culture could be common practice in another. For example, it is more common for grown children to live with their parents in some cultures than in others. Exploring these differences with interest and curiosity and without judgment is essential.

FOR GROUPS WITH YOUNG CHILDREN

Families in Recovery: What's True and What's Not?

- The truths and myths of recovery are important for children to recognize. Group leaders and parents can help present these ideas in ways that children can understand. For example, the parent or group leader can ask the child if s/he is worried that Daddy will drink again. Though the worksheet is written for older children or adults, the concepts are very appropriate for younger children, and the clinician should be able to express these concepts in ways that children can understand. Children need to know their parents are not using because of them, and they cannot make their parent use or not use.

Family Goals and Relapse Issues

- This worksheet generally applies to both adults and adult children.

FAMILIES IN RECOVERY: WHAT'S TRUE AND WHAT'S NOT?

Remember: Addiction is a Family Disease

This is true for almost all families in early recovery:

We don't trust you will stay clean and sober! Therefore, we are suspicious, worried, on alert, and hyper-vigilant.

In my family, this shows as: _____

_____

_____

_____

How I need to change my THINKING:

_____

_____

_____

_____

What I need to DO differently:

_____

_____

_____

_____

This is a common myth that families in early recovery believe:

If we are careful, you won't relapse!
   a. We believe we can cause you to use, and we really, really try to NOT make you use.
   b. We get afraid if you get angry or upset.
   c. We "walk on eggshells."

In my family, this shows as:

_____

_____

_____

_____

How I need to change my THINKING:

_____

_____

_____

_____

What I need to DO differently:

_____

_____

_____

_____

FAMILY GOALS AND RELAPSE ISSUES (Rawson et al., 2005)

Scale

1 = not at all, 5 = totally

Goals For Family Member(S) Beginning Stage (1–6 Weeks)

Make commitment to treatment	1	2	3	4	5
Recognize addiction as a medical condition (it's not about you)	1	2	3	4	5
Support discontinuation of drug and alcohol use (what does this mean?)	1	2	3	4	5
Look after yourself	1	2	3	4	5

What are three (or more) things you can do to move toward achieving these goals?

_____

_____

_____

_____

Goals For Family Member(S) Middle Stage (6–20 Weeks)

Learn to be supportive instead of co-addicted (no more unhealthy helping)	1	2	3	4	5
Begin finding ways to enrich own life	1	2	3	4	5
Practice healthy communication skills	1	2	3	4	5
In some situations, decide whether to recommit to the relationship (leave or trust)	1	2	3	4	5

What are three (or more) things you can do to move toward achieving these goals?

_____

_____

_____

_____

Goals For Family Member(S) Advanced Stage (20+ Weeks)

Learn to accept the limitations of living with an addiction	1	2	3	4	5
Develop an individual, healthy, balanced lifestyle	1	2	3	4	5
Monitor self for relapses	1	2	3	4	5
Be patient with the process of recovery	1	2	3	4	5

What are three (or more) things you can do to move toward achieving these goals?

_____

_____

_____

_____

Key Relapse Issues For Family Member(S)

Emotionally triggered by situations perceived as patient relapse	1	2	3	4	5
Fear of being alone	1	2	3	4	5
Lack of individual goals and interests	1	2	3	4	5
Unable to release responsibility for other	1	2	3	4	5

What are three other relapse issues for you?

_____

_____

_____

_____

What are three things you can do to lessen the power of these issues?

_____

_____

_____

_____

Families in Recovery II

Activity Title: Family Timeline

Activity Mode: Expressive Arts (drawing)

8

RATIONALE

Many families come to our groups after many years of struggling with a family member's addiction. They are often at the verge of or even well into hopelessness. One of the main purposes of this activity is to encourage hope in a future that is more positive than the present they have been living.

This activity is an expressive arts activity that helps families think about and express the timeline of their lives, including impactful experiences, both positive and negative. It also helps them develop a vision of a future that is not dictated by addiction.

Although it can be a powerful experience for an individual to see her or his own timeline, making a family timeline with family members can be even more powerful. Family members are given the opportunity to communicate about what experiences impacted each person and learn more about each other and their experiences. This activity can often trigger conversations about different individuals' experiences that can be profoundly moving.

OBJECTIVES

- To allow time for family members to work together
- To encourage family members to talk about the history of their own family
- To help family members think about what their family could look like in the future
- To help families make a plan to create the future they want for their family
- To encourage hope

MATERIALS

- Long pieces of paper (newsprint roll)
- Crayons, markers

Source: Chad Norris and Doug Moratz.

117

OPENING QUESTION SUGGESTIONS

- What is something you feel hopeful about (besides staying sober)?
- Where were you born, and where did you live for the first 5 years of your life?
- Do you have brothers and sisters? If so, where are they?

METHOD

1. Hand out large sheets of paper (they need to be wider than tall—preferably about 3 feet wide), one sheet per family. If large paper is not available, taping regular sheets will do.
2. Ask each family to draw a line across the middle of the sheet of paper from one end to the other.
3. Ask them to put a mark on the line ¾ of the way to the end of the line.
4. Describe the line as a "family timeline" with the mark they just put on the line being the present (today).
5. Ask them to decide what time frame the beginning of the line should represent (some may choose when children were born, when parents were born—it doesn't matter).
6. Ask them to divide the line up into segments representing years (could be every 5 years is segment, or every 10, or every 2, depending how many years the timeline represents).
7. Ask them to put major family events on the line. They can use words or drawings to represent these events. Events surrounding the family addiction should be included but should not be the only events on the line.
8. Allow about 15 minutes for this activity.
9. Then ask them to think about what they would like to put on the piece of the timeline that represents the future (after the mark ¾ down the page that represented today). Instruct them to take some time and draw/write words and images to represent their dreams for their family for the future.
10. Ask them to write down three things they can do to help ensure their dreams will become reality and write them down on their paper.
11. Bring group back to the circle and ask for volunteers to share their family timelines. Encourage them to share as much or as little as they are comfortable with sharing. It can be helpful to go around the circle and ask each person to share something about the experience even if they do not want to share their timelines.

GROUP DISCUSSION AND PROMPTS

- Ask the group what it was like to work with their families on this project. Did you have different ideas of what should be on the timeline? Did you find out about events that you didn't know about? Did it make you curious?
- Ask if there were any surprises as they worked on the project.
- Did everyone agree on what should be on the timeline?
- Ask what it was like to identify a vision for the future. What is the action plan you came up with?
- Did any emotions come up while you worked on this project?
- Are you more hopeful about the future than you were 3 months ago?
- Who tells stories in your family?

AVOIDING PITFALLS

- Group members often express discomfort at the idea of a drawing project. It can be helpful to emphasize that there is no way to do this activity wrong as long as they *do* the activity. Art can be involved as little or as much as they want.
- Some families have very difficult pasts, and it can be painful to review the history of painful events and trauma. Group members should not be forced to share more than they are comfortable with or even do more of the activity than they are comfortable with. There is a difference between refusing to do the activity because they don't want to, and balking because it stirs up pain. The group leaders must be adept at building relationships with group members so this can be assessed and addressed. Group members should always be encouraged to stay after group to process painful issues and emotions that arise at any time during group.

MODIFICATIONS

- Rather than creating the timeline as a family, it can be interesting to have each individual create his or her own timeline, then share it with other family members and notice what was the same and different on each timeline. This can help point out that different people can have very different experiences even though they are in the same family.
- This activity can be done without using words, only drawing. It can also be done using magazine pictures.

CULTURAL CONSIDERATIONS

- Group leaders should be cognizant of cultural differences with respect to what is appropriate to share and what is not. Acknowledging and respecting these differences is necessary.
- Groups will often include blended families. The timeline will be different for families that came together later in life. Group leaders can work with individuals and help them include individuals' lives and how they became a family.

FOR GROUPS WITH YOUNG CHILDREN

- This activity can be a wonderful chance for parents and children to create a timeline together. This can include sharing stories, answering questions, and creating a vision for the future from the child's perspective. Parents can learn a great deal from noticing what events stand out in a child's eyes. The timeline a child makes will likely be very different from the one a parent makes. It could be helpful for the parent to also create his/her own timeline (perhaps as homework) with his/her own goals and steps to achieving the goals.

9

Healing Families: What I Can Do

Activity Title: What I Can Do

Activity Mode: Psycho-educational (worksheet)

RATIONALE

Many of the people we work with in group want their families to change in some way. This activity helps individuals think concretely about what they would like to see change and develop an action plan for something they can do to improve how their families work.

The activity is ideal for working with group members who have families present, allowing them to work together to identify goals and a plan for follow-through. It can also, however, help group members without family members present think about what they would like to see change in their families.

In this exercise, we emphasize the idea that it is not possible to change other people, though it is very tempting to want to do so.

OBJECTIVES

- To nurture hope for individuals and families
- To help people recognize that the most effective way of encouraging change is to change yourself rather than change other people
- To develop at least one concrete action each person can take to help the family heal
- To help individuals recognize that by changing themselves, they can change their family system

MATERIALS

- Pens, paper
- What I Can Do: Change for Families worksheet

OPENING QUESTION SUGGESTIONS

- What is something you can change about yourself that will make your life easier (other than not using)?
- Have you ever made a change in your own life and then noticed that the people around you were changing as a result?

METHOD

1. Begin with a short discussion about what the group thinks about trying to change other people. Does this work? Is it possible? What are some times group members have tried to change someone else to make things better? Have the group complete the first two questions on the worksheet, then continue this conversation.

2. Explain that one of the goals of this activity is to help develop a feeling of hopefulness about the possibility of change in their families and provide some specific ideas of actions each member can take to help create this change. Emphasize that this activity will help all participants identify what they can do themselves rather than trying to get other people to change. Also point out that by changing yourself, you can change the family system (there is no way to avoid this!).

3. Give the group time to complete the remainder of the worksheet. Ideas for what they want more in their families could be peace, time together, fun times, dinners together, and the like. Families can work together, and group members can work in pairs if desired. (This can often spur further conversation and ideas.)

4. Allow time for the group to gather and share the commitments they are making, as well as who else they will share these commitments with. Let them know that the question of the week for next week will be about which of the commitments they worked on over the week. (Remember to use this as the question for the following week—as group leaders, we need to model keeping our own commitments.)

GROUP DISCUSSION AND PROMPTS

- A common experience for group members is that they want their families to trust them more quickly than is possible. This can be an interesting topic during this group—an example of group members making a change (not using) and then expecting others to change (increase trust) before the others are ready.
- Group leaders can introduce Al-Anon as a support group for family members who struggle with the issue of worrying about their using family members. Normalizing wanting their loved one to be sober and taking on more responsibility than is healthy can be helpful. It can be a relief for families to hear that this is a common experience for families struggling with substance abuse (and one of the issues that often arises in our groups). A conversation about what is healthy to take on and what is not can also be helpful, as well as emphasizing that staying sober is the using family member's responsibility, not the family's.

AVOIDING PITFALLS

- Group members who are not in contact with their families can have a difficult time with this exercise. Usually, group leaders can help these individuals identify some people in their lives that these ideas could be useful for. For example, they may have friendships that could be improved, or they may have a relationship with an employer they would like to improve. The concepts in this exercise can be translated to any relationship situation. For extremely isolated clients, even their relationships with treatment providers could be worked with. Brainstorming with clients as well as asking for suggestions from the group can be helpful. This situation can also serve as a guide for treatment in terms of developing a goal to build a sober support system.

CULTURAL CONSIDERATIONS

- There are no specific cultural issues that have arisen for us related to this exercise. Group leaders must always, however, be conscious of cultural issues and model respect, curiosity, and openness to how cultural issues may impact clients' experiences of this exercise.

FOR GROUPS WITH YOUNG CHILDREN

- This exercise is appropriate for groups with children. The idea of changing yourself (rather than changing someone else) is a concept that children can use throughout their lives.
- The conversations about what each (parent and child) can do to better their relationship can be helpful and productive.
- It should be pointed out to children that if they are ever in a dangerous situation, they need to tell someone (parent, teacher, etc.). The idea of not trying to change someone else does not mean they can't ask for help.

WHAT I CAN DO: CHANGE FOR FAMILIES

Most of us recognize that it is impossible to change someone else, yet most of us have tried to do that at some point in our lives! What was a time in your life you tried to change someone else?

_____

_____

_____

_____

Can you think of a time when you made a change yourself and then things got better with someone else? (In other words, you made a change rather than trying to make someone else change?)

_____

_____

_____

_____

WHAT I WILL DO

I can't change you, but I can change me.

Something I want more of for me and my family is: _____

_____

_____

_____

_____

To Get There

One thing I will start doing is _____

_____

_____

_____

One thing I will stop doing is _____

_____

_____

_____

Something I will do more of is _____

_____

_____

_____

Something I will do less of is _____

_____

_____

_____

A hope for my family is that _____

_____

_____

_____

To make this hope a reality, I will _____

_____

_____

_____

One person (in my family or not) I will share this hope and these commitments with is _____

_____

_____

_____

_____

What Could Be Better?

10

Activity Title: Making Group More Meaningful To Me—Fish Bowl; Group Rules

Activity Mode: Group Discussion

RATIONALE

This activity is particularly helpful in a group that is developmentally "young." This can happen if there are a number of new people entering the group or a number of the longer-term group members graduate from the group. As the makeup of the group changes in this way, the focus of the group often shifts to early developmental stages, which can include fighting for an identity. This can present as challenging group leaders, challenging other group members, and fighting group process.

The dynamics of each group are very different depending on the size, makeup, and developmental stage of the group. Our group is an "open" group, and members come and go regularly, so the makeup of the group is constantly changing. We sometimes experience periods when the group seems very young and unstable. This activity is especially helpful at these times and can help the group move toward a more highly bonded and committed state.

This activity encourages the group members to take responsibility for making group useful to them. Often, the more negative members of the group receive feedback from other group members that helps them feel an increasing sense of ownership of the group. In general, such feedback from group members has more impact than the same feedback coming from group leaders.

OBJECTIVES

- To help group members take responsibility for their experience in group
- To help group members recognize that they can get something positive from being in group and help them identify what that could be
- To facilitate group's cohesion and development

- To facilitate discussion about what group members want group to be for them
- To facilitate development of appropriate group rules
- To increase group members' feeling of ownership of the group

MATERIALS

- Large paper (newspaper roll), markers
- Group leader needs pad and paper to take notes

OPENING QUESTION SUGGESTIONS

- What is something you want to get out of group tonight?
- What is one thing you can do to make group more meaningful?

METHOD

1. Group leaders talk about their experience of noticing unrest in the group. Group leaders can speak with other group leaders about this in front of the group.

2. Group leader can suggest the group have a conversation about what they want group to be like.

3. Arrange a "fishbowl" in which group members sit in a circle on the inside and group leaders sit on the outside. Tell the group the group leaders will listen only, unless they are asked specific questions (there will be time for group leaders to talk afterward).

4. Group leaders can pass in written questions. Some ideas for questions are: What would it take for the group to be more invested in the time spent here? What kinds of activities would the group like to see? What about the group participating in developing activities? Questions will be based on what has been going on in group.

5. After group discussion has ended (10 to 15 minutes generally, though if the discussion is going well and it seems helpful to continue, group leaders can decide to do so and facilitate the next part of this session—developing group rules—next week), group sits together again with group leaders, and group leaders talk with group about what they heard. Let the group know that group leaders will take group's suggestions seriously (have written them down) and will bring typed suggestions next week as we further the conversation.

6. Depending on the time left after discussion, there may be time to have the group develop group rules. If not, continue with this part of the activity for next week.

7. Group leaders facilitate brainstorming a list of group rules (use the board). Point out that most rules should be based on safety and respect. Have no limits to the brainstorming section—write down everything on the board. The next step will be to have the group identify the rules they need. Group leaders should be sure that issues such

as timeliness, turning cell phones off, and any other rules they see as necessary be part of the list.

8. Once final rules are identified, group can break into small groups, each take a rule, and write it on a strip of newsprint (be creative). Separate strips can then be glued/taped onto a large sheet and hung in the group room. Group rules should be reviewed regularly.

9. Ask group for feedback on this activity, and ask what impact they feel that doing this will have on the group.

10. Group closing.

GROUP DISCUSSION AND PROMPTS

- In our experience, the group will be aware of and acknowledge when the group is not a strong group. It is rare, however, that there are not at least a few stable, committed group members, and these members often feel frustrated that other group members are not "serious." This activity can help these group members find their voice and give them an opportunity to express these frustrations, as well as their desire to have support for their recovery. Group leaders can acknowledge these statements as the group members standing up for their recovery.
- This activity can be run as a 1- or 2-week topic. If there is a lively discussion in the fishbowl section, group leaders can give homework to the group of considering rules they feel would be helpful and lead the group rules section of this topic the following week.
- Normalizing different group stages can be helpful, and presenting this as an opportunity for group members to create what they want can help group members be invested in participating in this activity.
- If the group is struggling as the fishbowl activity begins, group leaders' passed-in questions (see above for suggestions) can help start the conversation.

AVOIDING PITFALLS

- This can be an intimidating group to lead, as generally it is run at a time when the group is not strong—sometimes frustrated and angry. In our experience, the way to have group leaders survive this developmental process is to make sure that group leaders are able to support each other throughout the process. That means that group leaders sit in areas of the room where they have easy access to eye contact with each other, and they talk out loud with each other about any discomfort they are feeling (including not being sure where to go next) and concern that the group moves forward.
- Group members can become frustrated if they feel their concerns are not heard and responded to. It is helpful for group leaders to let the group know that they function within certain program restraints and that they have the authority to facilitate some changes and not others. The group must know which of their requests can be implemented and which cannot.
- Although not all changes the group wants will be approved, the changes that the group suggests that are approved must be adopted.
- In our group, we have a treatment team that makes final decisions. It was necessary to let the group know we will take their suggestions back to the treatment team and let

them know the following week whether we can make the changes they are requesting. Again, it is essential that this happens the next week and group leaders follow up on their commitment.

MODIFICATIONS

- This can be a 1- or 2-week activity.
- It is possible to reverse the fishbowl once the group has finished its processing and switch places—have the group leaders sit in the center and talk about their experience of listening to the group while the group listens (and can pass in questions). This can be especially helpful, as group leaders can then point out their experience of group members standing up for their recovery (or their disappointment that they did not).

CULTURAL CONSIDERATIONS

- In most groups, there are members who are vocal and members who are not vocal. Group leaders should be aware that speaking in group can be intimidating and provide support and encouragement to those that are not speaking. One way of doing this is to ask a more quiet group member if he/she would like to talk about what he/she is thinking. This approach allows the group member to refuse, but our experience is that often this approach gives the quiet group member an opening to share what she or he is experiencing.

FOR GROUPS WITH YOUNG CHILDREN

- The fishbowl can be used to allow the children in the group to express their thoughts about what would make group better. In groups that include children regularly, one part of the fishbowl can be used to have the children in the center, with the adults and group leaders listening and passing in questions. There may be a need for more questions to help direct the children's discussions. Questions such as "What could make group better for you?", "What can you do in group to get more out of it?", "What are some activities you would like to do with your parents in group?", or "What would you like to learn in group that would help your family?" could all be useful.

Rebuilding Trust

11

Activity Title: Rebuilding Trust Discussion and Worksheet

Activity Mode: Psycho-educational (worksheet)

RATIONALE

Almost universally, people who present for substance abuse treatment have families who have lost trust in them. This can be a painful subject that can provoke strong emotions.

It can be helpful for group members and family members to process their own issues and hear other families process their issues related to trust. One benefit is that families often recognize that these experiences, though painful, are very common. We often hear group members and family members express surprise and relief as they recognize that they are not alone with these experiences.

In most groups, there will be individuals and families in various stages of recovery. It can be helpful for families that are new to treatment to hear stories from families that have been working on these issues for some time and have experienced some success. This experience can be the beginning of rekindling hope.

OBJECTIVES

- To help families understand that it is not uncommon for trust to be lost when substance abuse is involved
- To help families understand that trust can be rebuilt
- To help families think about how to rebuild trust in their families
- To help families see building trust as an ongoing process rather than a single event
- To help families have hope that trust can be rebuilt

MATERIALS

- Rebuilding Trust worksheet
- Paper, pens, pencils

OPENING QUESTION SUGGESTIONS

- What is something you think about trust?
- Who is someone in your life that you trust? Why?
- How do you know when you can trust someone?

METHOD

1. Introduce the topic of trust. Ask the group for their experiences with trust within their families.

2. Normalize problems with trust in families with substance abuse. Ask the group if there is anyone present who has NOT experienced problems with losing trust when they were in their addiction (most likely, no one will raise his or her hand).

3. Present rebuilding trust as a gradual process rather than a single event (not like turning a light switch on). Normalize difficulties.

4. Ask about differences between trust and forgiveness. (You can forgive someone and still not trust him or her. Forgiveness is a decision on the part of the offended party. Rebuilt trust develops based on the offender's acts.)

5. How do you know you can trust someone?

6. What can you do to rebuild trust (see worksheet)?

7. Hand out worksheet and give time for group members to complete it.

8. Have everyone regroup, and ask group members what it is like to think about these issues. Encourage all participants to share at least one thing they identified that they can do to rebuild trust.

GROUP DISCUSSION AND PROMPTS

- Many of our clients are impatient for their families and supports to trust them again. A conversation about patience can be helpful, as can normalizing both the desire to have trust quickly (I've changed, why don't they trust me?) and the need for time (it took years to lose trust, and it could take years to get it back). This can be especially effective if group members can talk about their experience of being patient and of seeing that trust can be gradually regained.
- Acknowledging and normalizing the pain involved in recognizing that group members have lost trust can be helpful, as is acknowledging the family's pain at the recognition that they can no longer trust their family member.
- Encouraging group members and family members to share their experiences with trust (or lack of trust) can be helpful. Often, these difficult emotional experiences can be understood better coming from a family other than their own family. We find that it can be particularly helpful for clients to hear other families talk about the concept of being patient as families redevelop trust.
- It is important to have each person in the group identify one concrete action step he or she can take to move in the direction of increasing trust.
- It can also be helpful to ask the group whether this exercise helps them feel more hopeful that trust can be rebuilt in relationships.

AVOIDING PITFALLS

- There may be people in group who no longer have contact with their families and who have no plan to reconnect with them. These people can be encouraged to think about trust with regard to the people that are in their lives currently.
- Again, this activity can bring up strong emotions, so group leaders must be sensitive and aware of strong responses and offer support.
- Group leaders should be conscious that this process does not fall into a complaint session about families who have not regained trust. The idea of patience (it takes time to rebuild trust) can be particularly useful, and generally there will be group members present who have experienced this.

MODIFICATIONS

- If there is a large number of family members in the group, it can be interesting to divide the discussion about trust into three sections:
  o Section 1—Conversation between family members only (family members can bring their chairs to the center of the circle)
  o Section 2—Conversation between family members in recovery only
  o Section 3—Conversation with everyone together (What was it like to hear the other group talking about this issue?)

CULTURAL CONSIDERATIONS

- The concept of trust can vary between cultures, and encouraging an open conversation about these differences can be helpful. Ask what it was like for the person struggling with addiction in her or his culture. What was the person's relationship with his or her parents like? Expressing respect, openness, and interest is essential.

FOR GROUPS WITH YOUNG CHILDREN

- This is a topic that many children can relate to, and it can be very appropriate for groups with young children. Children can be encouraged to talk about trusting (or not trusting) their parents. This can be an emotional conversation, and group leaders must (as always) be aware and sensitive to this.

REBUILDING TRUST

Discussion and Worksheet

The following are some suggestions for ways to work toward rebuilding trust in relationships. Read each suggestion, and think of ways you can apply it to the relationships in your own life.

Keep My Word

_____

_____

_____

_____

Be Consistent

_____

_____

_____

_____

Be Considerate

_____

_____

_____

_____

Keep Trying—Be Patient

_____

_____

_____

_____

Listen

_____

_____

_____

_____

Apologize

_____

_____

_____

_____

Other

_____

_____

_____

_____

Who are the people I want most to trust me again? (Or who are the people I most want to trust again?)

_____

_____

_____

_____

What are some things I can do to move in this direction?

_____

_____

_____

_____

12

Building Lifelong, Healthy Supports I

Activity Title: Parts of My Life

Activity Mode: Expressive Arts, Psycho-educational

RATIONALE

By the time many of our group members arrive at our group, they have burned many bridges in their lives and often have very few supports. The supports they have are often unhealthy. Many group members struggle with letting go of people in their lives that do not support their recovery. For some, that means almost everyone they know. This can be a painful and difficult process.

This activity acknowledges the pain of this situation and works to develop hope in the possibility of finding and developing new, healthy relationships that support one's recovery. It also provides an opportunity for group leaders to introduce the benefits of attending AA/NA and obtaining a sponsor. Finally, this activity serves to help people identify various pieces of their lives and think about whether there are some pieces they would like to play a larger part or some they would like to be smaller.

This activity can also help groups that are struggling with group members that are not bringing in family members. Since during this activity, group members identify specific individuals who are supportive in their lives, group leaders can then ask the group member if he or she would be willing to bring this person to group. If the group member is not willing, that can open up a discussion of the group members' concerns.

This activity includes an expressive arts activity that helps group members identify areas of their lives they have support and areas in which they need to build support.

OBJECTIVES

- To acknowledge pain of having burned bridges in one's life and subsequent losses
- To acknowledge pain of not being able to spend time with old "friends"
- To acknowledge loneliness and introduce the HALT (hungry, angry, lonely, tired) concept from AA
- To inspire hope in the possibility of creating healthy, supportive relationships
- To identify areas in which group members would like to build more support
- To give information about community supports available

MATERIALS

- Parts of My Life worksheet (though larger sheets of paper can be used)
- Pens, crayons, markers
- Information about recovery-related support groups in the area (can be obtained online)

OPENING QUESTION SUGGESTIONS

- Who is in your life now that is a positive, healthy support for you?
- What is something you do these days that is healthy fun?

METHOD

1. Open the group by asking group members who in their lives is supportive (this could be the opening question). This can lead to a general discussion of people with whom group members have had to cut off communication. Allow some time for group members to share their experiences.

   a. Ask if one of the results of these changes is dealing with feelings of loneliness and how group members are coping with these feelings. Ask if someone in the group is familiar with HALT, an AA concept that cautions against becoming hungry, angry, lonely, or tired. Facilitate processing of people's experiences with these emotions and their impact on recovery.

2. Facilitate expressive arts exercise. Let the group know we will be doing an exercise that will help identify supports and positive activities in people's lives. Hand out Parts of My Life worksheet and something to write with.

   a. On one side of the paper, ask the group to take some time to identify and write down several areas that are important in their lives. Examples could be friends, work, recovery, spirituality, family, and so forth.

   b. Then, ask them to draw a large circle (as large as possible) on the other side of the paper. Divide the circle into sections (like a pie), with one section for each area they identified. Sections can be sized according to the amount of time they take. Label each section (family, work, recovery, etc.). Labels can be written in words or drawings.

   i. Ask one or two people to share what they have drawn.

   c. In each section, write details about what is supportive or helpful. Write names of specific people and specific activities. For example, go to church, go to AA, specific names of friends and family members, specific activities. In the "work" section, it may be "the routine" that is helpful.

   i. Ask one or two people to share what they have drawn/written.

   d. Ask group members to look at their "pie" and consider if they would like to add other pieces or take away some of the pieces. For example, a group member may want to put more emphasis on spirituality in his or her life. Ask each group member to write down something he or she would like to add to (or take away from) the pie, and one thing he or she can do to move in this direction.

   i. Ask one or two people to share what they wrote.

   ii. Ask if anyone wants to redraw his or her pie, and hand out paper for them to do so.

    e. Ask group members to identify positive supports specifically for their recovery. Normalize the fact that many group members may not have developed those supports yet—that is okay; that is one of the points of this activity.

    i. Ask one or two people to share who they have identified.

3. If AA/NA has not come up in the conversation already, group leaders should ask the group if there are people there who attend AA/NA and/or have a sponsor. Ask those group members to share their experiences. Group leaders should also point out that AA/NA connections are lifelong, whereas time in this group is limited. Also, though it is easy to feel strong connections with fellow group members, it is important to remember that everyone in the group is in early recovery. (This can lead to a conversation about potential risks of spending time with other group members, which can be a sensitive subject.) Group leader can provide group with information about AA/NA and other recovery-support-related meetings in the area (access online).

GROUP DISCUSSION AND PROMPTS

- Group members who have family members or supports with them in group can be encouraged to let them know what it means to them to have their support. This often leads to feedback from other group members to the family members, acknowledging their support for the group member. Many group members will not have that kind of support.
- The idea of avoiding close relationships between group members can be a difficult subject. We have a group rule that disallows sexual relationships between group members unless they had that relationship before beginning group. (We allow couples in our group.) On the one hand, it is helpful and we want to encourage group members supporting each other. On the other hand, we have seen group members relapse together many times. This is an ongoing conversation and balance and is something our treatment team continues to deal with. There is not a black-and-white solution.
- There will likely be varying experiences with AA/NA, some positive and some negative. Group leaders can encourage group members to try it again and remind them that often, something that is not helpful at one point in our lives can be helpful at another point.
- One barrier to people attending AA/NA is fear of going into a group of strangers alone. Often, group members will offer to accompany a new person to a meeting. This can be very helpful.

AVOIDING PITFALLS

- Group members who have limited supports and feel badly about their lives can have a negative reaction to this activity. In such situations, group leaders can suggest that the "pie" represent what they would like their life to look like rather than what it is now. Normalizing these difficulties is essential. There will likely be several people in group who can relate to this experience.
- There often are people in group that have had negative experiences in AA/NA. Accepting that experience and encouraging input from others who have had more positive experiences can help with this. Also (as above), remind the group that something that doesn't work at one point can work at another (recovery, for example). Asking the group who has had that experience can also be helpful.
- In our area, there are only a few NA meetings and many more AA meetings. One common objection to attending AA is that people with addiction have felt they were not welcome. Again, gaining feedback from the group can be helpful in addressing this issue. Probably the most effective response is to acknowledge that difficult and painful experience and encourage the group member to be open to trying again. Every group is different, and certainly one negative experience does not mean that all AA/NA is that

way. We always encourage group members to attend at least a few meetings before making an evaluation.

- Group members are often hesitant to bring family members to group. Acknowledging these fears, asking the group member to talk about what their fears are, and encouraging input from group members and family members who have had a positive experience in group can help with this. Our experience is that it takes time for group members with this stance to be more open. We often ask for permission to bring up the issue again at a later time with the group member.

MODIFICATIONS

- This exercise can be facilitated in a two-step format, with the first being drawing a pie showing the way things are now and the second drawing the pie the way they would like it to be. Follow up with steps they need to take to move toward life as they would like it to be.
- Families can draw a joint Parts of Our Family drawing using the same directions but focusing on the family rather than the individual. It can be helpful to do both—each person in the family creating her or his own individual drawing and then working together to create the family drawing. Again, follow up with ideas of how to move from where they are to where they would like to be.

CULTURAL CONSIDERATIONS

- Group members often bring up going to church as a support for their recovery and as a replacement for AA/NA. We always support this and certainly acknowledge a strong spiritual path as an extremely important piece of a healthy life. We also ask whether there is someone with whom the group member can share issues related to her or his recovery, as we see this as an important piece of an ongoing, stable recovery.
- Families can be frustrated with the amount of time the group member spends in recovery-related activities. This can be a helpful topic in group, both to normalize as well as for group leaders to emphasize the importance of this recovery-related time (particularly in early recovery) and to offer support for family members for whom this is difficult. This can be a particular problem in families that had infidelity issues and should be processed in family sessions with a counselor.

FOR GROUPS WITH YOUNG CHILDREN

- This can be a helpful activity for families with young children. Children and adults can work together to create a family pie that shows the parts of their family currently and then one that shows how they would like it to be. They can then identify steps they can take to move toward how they would like it to be. For example, children may want more playtime with their parents. Or they may want more TV time! This will be a chance for parents to set boundaries (they do not need to accept everything their children want!) while acknowledging their children's desires. This can open the door to wonderful conversations between parents and children, with the possibility of moving toward a family-focused life together. Examples of what could come out of this are a family game night or a time when parents have one-on-one time with children.

PARTS OF MY LIFE

List parts of my life:

1. _____Recovery _____   6. _____
2. _____   7. _____
3. _____   8. _____
4. _____   9. _____
5. _____   10. _____

Some thing(s) I would like to add to my life is/are:

_____

_____

_____

_____

One thing I can do to move toward having more of the above in my life is:

_____

_____

_____

_____

Some thing(s) I would like to have LESS of in my life is/are:

_____

_____

_____

_____

One thing I can do to move toward having less of the above in my life is:

_____

_____

_____

_____

One person (or more) I can ask to come to group with me is:

_____

_____

_____

Building Lifelong, Healthy Supports II

13

Activity Title: AA/NA/Al-Anon

Activity Mode: Speaker

RATIONALE

We find that group members often become very connected with the group and have a difficult time connecting with other supports. We emphasize that our group is time limited, and it is essential to build other sober supports that will stay with them their entire lives. This can be difficult for some group members.

The purpose of this group is for group and family members to hear the story of someone who has been in recovery for a long time and has built these sober supports. In our area, AA and NA are the only recovery-specific groups available, so most of the people we bring in have connections with these groups. In other areas, there would be other options. We also bring in people from Al-Anon who are willing to share their stories.

This can be an informal, inspirational evening in which many of group members' concerns about attending AA/NA and Al-Anon can be addressed. The format of this group is conversational—generally, guests share their stories, and the rest of the evening is spent with questions and answers. The evening can be inspiring and hopeful.

OBJECTIVES

- To inspire and build hope in recovery
- To provide information about supports available in the community
- To acknowledge nervousness and concerns about these supports and address the concerns
- To emphasize importance of developing lifelong supports and point out that this group is time limited

MATERIALS

- Information about community supports (AA/NA/Al-Anon and other available supports)

OPENING QUESTION SUGGESTIONS

- What is one thing you do (besides group) to support your recovery?
- What is one change your family has made as a result of your recovery?

METHOD

1. Let the group know that we have a guest (or guests) who will be telling the story of his/her recovery. Introduce the guest (or guests). Work with the guests to find out how they would like to be introduced.

   a. We encourage group leaders to be familiar with resources available in their community.

2. Ideally, have one guest from AA/NA or other substance-recovery-related group and one guest from Al-Anon.

3. Allow time for the guest to share his/her story and then time for questions.

GROUP DISCUSSION AND PROMPTS

- Generally, this group runs basically on its own. Group leaders can be prepared with some questions in case the group is shy to begin with.

AVOIDING PITFALLS

- Have a clear understanding of the speaker's agenda and make sure there can be time for the group to ask questions.
- Acknowledging that the group has concerns (sometimes negative experiences) can be helpful and can lead to a discussion of those concerns. It can be helpful to let the speaker know that these concerns may arise.

MODIFICATIONS

- This activity can be divided into two weeks, with one week for an AA/NA speaker and another for an Al-Anon speaker.

CULTURAL CONSIDERATIONS

- Although we would not bring in a speaker with a particular religious orientation, many group members will have religious views and may see their church as their main support for their recovery. This can be an excellent topic to bring up with the speaker. Respect for all religious perspectives is, of course, essential.

FOR GROUPS WITH YOUNG CHILDREN

- Many areas have Alateen groups. Having a speaker from that group could be helpful and interesting.
- At times, the stories speakers tell can be emotionally difficult. Parent should be aware of this and able to decide whether children should hear these stories. Parents and group leaders can talk to speakers to determine appropriateness for children.

14

Spirituality

Activity Title: Spirituality and My Family

Activity Mode: Expressive Arts (drawing)

RATIONALE

Talking about spirituality can be difficult. People often have strong ideas and opinions. It can be a challenge to introduce this topic and guide the activity in a way that acknowledges and supports everyone's experience of spirituality.

Spirituality is also an important piece of recovery, and many people in our groups say that their growing spirituality is one of if not the main foundation of their recovery.

This main goal of this group is to introduce the concept of spirituality as an important piece of recovery and encourage a supportive environment for people to share their own experiences of spirituality. The goal is *not* to advocate for one religious or spiritual path or another, and the group leaders should make this clear.

This exercise begins with a discussion, followed by the expressive arts activity, followed by sharing.

OBJECTIVES

- To introduce the idea of spirituality as an important piece of recovery
- To help group members consider the development of their spirituality as it relates to their family experience and explore the spirituality of their families
- To normalize the fact that group members may have different spiritual paths
- To allow an opportunity for group members to share their own experiences with spirituality and to hear others' experiences
- To help group members explore the relationship between spirituality and their recovery using an expressive arts exercise
- To normalize struggles with spirituality

MATERIALS

- Art materials—paper, markers, crayons, paint, etc. Magazines, glue, and scissors could also be available for group members to make a collage.

OPENING QUESTION SUGGESTIONS

- Have you had someone in your life you consider a spiritual teacher?
- In a few words, say how spirituality has impacted your life and your recovery.

METHOD

1. Begin with a discussion of spirituality in your life, in your family's life, in your recovery. See below for more specific suggestions of how to encourage the discussion. This discussion should take place while the group is still in a circle.

2. Once the group has processed for some time and the group leader feels there has been enough of an introduction of the topic of spirituality, let the group know we will be doing an expressive arts exercise on the topic of spirituality.

3. Pass out paper and art materials and give the following instruction: "There is absolutely no right or wrong way to do this activity. I want you to sit in silence for a moment, close your eyes if you like, and take a few deep breaths, feel your body relax, and then just sit in the silence. Be with the feelings that have come up in you as a result of the last few minutes of talk about spirituality in your and your family's life, about spirituality and your recovery. Notice the feelings—they may be soothing, calm, restorative feelings—or maybe some other feelings came up in you, confusion or anxiousness, even anger. We all have many feelings about our spirituality, and that is fine. Some may be clear about our spiritual path, some may not. Maybe we have a family in which everyone is on the same spiritual path; maybe people in our family have different spiritual experiences. It doesn't matter. Right now, we will put our awareness on our own spirituality. How it feels, where it is in our lives, how we feel with it—what it is to us. This is what we will draw. No right or wrong. With no talking, just start drawing any image that has come to you. It doesn't have to look like anything, so don't worry. Just draw what your spirituality means to you, in your family and in your recovery. Use colors, words, images—whatever you want. If nothing comes to mind, just draw. There is no right or wrong.

4. Give the group about 10 minutes (more or less as seems right) to finish their drawing. Encourage them to do this exercise in silence.

5. Bring the group back together and encourage them to share their drawings and their experiences creating their drawings. Ask them if this brings to mind anything they would like to work on or change about where spirituality is in their lives.

GROUP DISCUSSION AND PROMPTS

- Often, the group discussion around this topic flows easily. Group leaders should make sure the following ideas are processed:
  - What was spirituality like in your family growing up? Has that changed for you?
  - How do you support the spiritual development of your children?
  - How has your spirituality impacted your recovery?
- When sharing drawings, group leaders can ask:
  - What was the experience of making this drawing like for you?
  - How does the drawing relate to your spirituality?
  - Does this drawing or exercise make you think of anything you would like to work on changing with respect to spirituality in your life or in your family's life?
  - Is there something you want to do to strengthen your spirituality?

AVOIDING PITFALLS

- It is essential that group leaders model a supportive stance for any spiritual paths that are presented. It can be helpful to let the group know that this is not about pushing any particular religion or belief system but about encouraging the development of a spiritual life (just like we encourage healthy physical, mental, and emotional lives). Group leaders can remind group members that no one should try to convince anyone else that their way is "right."
- Distinguishing between religion and spirituality can be helpful to do at the beginning of group, and group leaders can come back to this distinction if group members move toward pushing their own religion.
- Presenting this as a potentially volatile topic can be helpful. Again, emphasize that this group is not meant to convince people to choose one spiritual or religious path over another one. This information can be given numerous times during the discussion. Group leaders can also let the group know that they will not allow arguments about religious differences in this group.

MODIFICATIONS

- The expressive arts activity can be done by individuals, or families can work together on a family spiritual drawing...OR
- Individuals can create one drawing and share it with the group, and then families can create another one together.

CULTURAL CONSIDERATIONS

- There will likely be group members from various religious backgrounds in the group, some of whom have very strong opinions about their paths. Group leaders must be proactive by letting the group know that arguing about religion is not what this group is about. At the same time, group leaders can acknowledge the fact that people do have strong and passionate opinions about this topic, and that is a good thing. It is just that this group is not the place to talk about that. This is the time and place for supporting everyone's spirituality.

• There may be some individuals who are nervous about sharing their spiritual beliefs— often people who have religious beliefs that are in the extreme minority. It is important that any such concerns be acknowledged (if they are shared) and that the safety of group members is respected. If a group member is not comfortable sharing his or her spiritual beliefs, this must be respected. At the same time, group leaders should model respect and curiosity and an openness to learning about another's experiences.

FOR GROUPS WITH YOUNG CHILDREN

• This can be an excellent activity to use with families with young children. Children generally have no fear of drawing (as many adults have), which can make this activity flow very smoothly. Also, encouraging parents to talk with their children about spirituality can be helpful. Children often have ideas about spirituality their parents do not even know they have—these can be fantastic conversations. Also, creating a drawing together of how spirituality fits into their family's life can bring in aspects of life that many families have never talked about.

15 Developing Healthy Rules

Activity Title: Rules in My Family

Activity Mode: Psycho-educational (worksheet)

RATIONALE

This activity focuses on rules in families—both spoken and unspoken rules. All families have rules, and in general, families that function well generally have clear rules. As addiction takes over in a family, the rules relating to structure of the family (specific times for meals, time to get up in the morning, bedtime, having family members eat meals together) often fall apart. At the same time, unspoken rules often become more powerful (don't talk about Dad's drinking, don't come out of your bedroom when Mom and Dad fight, do defend your family at all costs).

The purpose of this activity is to bring these patterns to light, help families identify current rules that are functioning in their family, and evaluate those rules, with the possibility of changing the rules in the family. We emphasize that as a family grows in stability and health, many things will change, and developing a healthy structure and clear rules is a sign of positive changes.

OBJECTIVES

- To help group members recognize spoken and unspoken rules in their families
- To help group members acknowledge how the rules have changed with the onset of addiction in their families (though some group members will not have experienced living in a family without addiction)
- To help group members recognize the importance of healthy structure and rules in their families and create healthy rules
- To help group members recognize the power of unspoken rules, identify them in their families, and bring them into the open, with the hope of either changing the rule or consciously continuing the rule
- To help group members identify positive changes their families are making as a result of their work

MATERIALS

- Rules in My Family worksheet
- Pens, pencils

146

OPENING QUESTION SUGGESTIONS

- What is a rule in your family that was hard for you to follow?
- What is a rule that was never spoken out loud, but everyone in the family knew about?
- What does your family do at mealtimes? Was this what you did when you were growing up?

METHOD

1. Encourage discussion about rules in families. Have this discussion lead into the value of structure in families.

2. Hand out the worksheet and have the group complete one section at a time, with time for sharing after each section.

3. Ask the group if they have noticed positive changes in the realm of rules or structure since they have moved into recovery. Acknowledge that these changes often take hard work, so it is okay if they have not noticed them.

4. At the end of group, go around the circle asking group members something they learned (or thought about differently) as a result of this group and something they will change as a result of this.

GROUP DISCUSSION AND PROMPTS

- Are the rules in your current family similar to the rules in the family you grew up in? What do you think about this?
- What happened to the rules in your family when substance abuse "took over"?
- Many families have powerful though unspoken rules to not talk about family secrets. Sometimes these secrets are related to substance abuse, but not always. Has anyone experienced this who is willing to talk about it?
- Normalize that these subjects are often very difficult to talk about. It is important that group leaders offer an opportunity but no pressure.
- Talk about the idea that structure and routine help children feel safe, and encourage parents in the group to talk about what structure and routine they have in their families. Also make the point that in families with substance abuse, healthy routines often change to unhealthy routines (Mom starts drinking at 3:00), and these new routines can start to feel "normal." This is an opportunity to take a look at routines and structure in our families, evaluate them, and consider what could be improved.
- Parents' guilt at not providing a safe environment should be acknowledged, as well as the fact that though they cannot change the past, they can make changes starting immediately that will help their children in the long term.
- Generally, people will notice that the rules and structure they were raised with have a significant impact on the rules and structure they create in their current families. This can be an interesting topic of discussion, again leading to an evaluation of what we are doing and the possibility of making positive changes.

AVOIDING PITFALLS

- Group leaders should be prepared to give examples of rules and ask questions that will lead group members to be able to identify the rules in their own families.
- Acknowledging and supporting parents who have not provided a safe environment for their children can be helpful. In our groups, there are generally several members who have lost custody of their children, and this can be a painful topic.

MODIFICATIONS

- An expressive arts activity can be added to this chapter in which group members draw a frame to represent the structure of their family and draw images to represent the rules and structure they have in their family (Section III of the worksheet). This is an excellent way to reinforce the idea that structure and boundaries (the frame) allow children to feel safe.
- There may be group members (such as young adults) that are currently living in their family of origin. In such cases, they can complete the worksheet by answering the questions based on how their family was when they were children compared to how the family is now.

CULTURAL CONSIDERATIONS

- Different cultures will have different cultural norms for rules in families. A group conversation about what is normal in various cultures can be interesting and helpful in furthering the concept that there is no one way to become a healthier family.

FOR GROUPS WITH YOUNG CHILDREN

- Talking about rules with children can be helpful and enlightening. Children will often bring up unspoken rules that parents were not aware the children were following. This can be a good opportunity for parents to give permission to children to not continue with the rules. For example, children may be following the rule that "we don't talk about Daddy's drug use" even though they were never told (in words) to do that. In such situations, group leaders should help parents respond without becoming defensive and focus on the fact that having this information provides parents with the opportunity to make changes that will have a positive impact on their children.
- For families that have had little structure, this can be a good opportunity to work toward developing rules (and routines) that will strengthen the health of the family. Group leaders can work with parents and children together to develop a set of family rules that will work. The set of rules can be a work in progress that can be reassessed as the family lives with them. Group leaders can encourage families to continue this process in their family therapy sessions (separate from group).

RULES IN MY FAMILY

I. Spoken Rules

    a. What are some of the spoken rules in your family as you were growing up?

_____

_____

_____

_____

    b. What are some of the spoken rules in your current family?

_____

_____

_____

_____

    c. How did/does addiction impact these rules?

_____

_____

_____

_____

    d. What are some rules you may think of bringing into your family that would help increase the healthy structure of your family?

_____

_____

_____

_____

II. Unspoken Rules

    a. What are some of the unspoken rules in your family as you were growing up?

_____

_____

_____

_____

b. What are some of the unspoken rules in your current family?

_____

_____

_____

_____

c. How did/does addiction impact these rules?

_____

_____

_____

_____

d. What are some unspoken rules you may think of talking about with your family? Are these rules you want to keep or change?

_____

_____

_____

_____

III. Structure for My Family

a. What are some things you do to create a more healthy structure for your family? Examples of this might be: having dinner together, having a set bedtime, picking up the kids at school every day, having an after-school snack, going to church every Sunday, having a family movie night, etc.

_____

_____

_____

_____

b. What ideas have you thought of during this activity that would help create more of a healthy structure for your family?

_____

_____

_____

_____

Curriculum Section V:
Anger

1

Anger in My Family

Activity Title: What I Learned About Anger and What I Want to Teach My Children Worksheet

Activity Mode: Psycho-education (worksheet)

RATIONALE

During the treatment process, most families recognize that there are relatively stable patterns of interactions that occur in multiple generations. Substance abuse is one of these common patterns, as is how anger is managed and expressed in families. The process of helping individuals identify these patterns (both healthy and unhealthy patterns) and determine which of these patterns they want to support and continue and which they want to change can be an important part of treatment.

It is common for anger in families to be one pattern that many group members want to change. This exercise can help individuals learn more about where they learned their anger responses and identify other responses that could work more effectively.

This exercise is about identifying patterns and developing a vision of changing those patterns. It is particularly effective in families with children, as the vision is for changing children's experiences, resulting in a positive change in their lives. This can be a hopeful experience for parents.

Many group members carry a great deal of guilt and shame about how their anger has hurt others. Hearing others talk about their experiences can normalize these painful experiences. We hear many individuals, both addicts and family members, express relief when they realize, as others share their experiences, that they are not alone.

OBJECTIVES

- To continue to facilitate an open discussion about anger
- To normalize difficulties managing anger appropriately
- To develop hope in the idea of changing patterns in families
- To normalize guilt and shame about how people have managed anger inappropriately in the past
- To help people see how what they learned as children impacts how they experience anger today

MATERIALS

- What I Learned About Anger and What I Want to Teach My Children worksheet
- Pens and pencils

OPENING QUESTION SUGGESTIONS

- Do you think you handle anger in a similar way to someone else in your family?
- What is one thing you learned about anger from growing up in your family?

METHOD

1. Hand out What I Learned About Anger and What I Want to Teach My Children worksheet.

2. Talk about cycles in families and about how children learn what they live. Talk about the fact that what children experience in their lives becomes "normal" to them. Ask the group to talk about how they have experienced this in their own lives.

3. Have group draw or write their responses to the worksheet.

4. Give time to share responses.

GROUP DISCUSSION AND PROMPTS

- The group can spend some time talking about patterns and cycles in families. Asking the group to share what has been passed down from one generation to another in their families can bring up many different topics. Examples could include cycles of substance use, violence, family size, creativity, music, and so forth. Not all these patterns and cycles will be negative.
- Asking the group what they learned about anger as children can be helpful. Group leaders can take these experiences and help group members be more aware of experiences their children may be having.
- Talking with the group about how experiences in children's lives become "normal" for them can also be useful. Many group members will be able to identify experiences they had that seemed normal at the time but in retrospect are not. Examples of this are living with parents that abuse substances or living in dangerous situations. Group leaders can normalize the tendency to fall into these same patterns and cycles and acknowledge the courage and strength group members are displaying to address and work to change them.

AVOIDING PITFALLS

- This topic can bring up strong emotions. Many of our clients have trauma from anger-related incidents. Group leaders must be sensitive and aware of this possibility, and should encourage group members to bring up issues in the group if possible. Speaking about these issues in group may be difficult for individuals experiencing deeper trauma. In such situations, the group member can be encouraged to speak individually with group leaders until they feel more comfortable. Group leaders should emphasize that these emotions can sometimes trigger desire to use and emphasize the importance of addressing them as they arise.

- Sometimes this topic can bring up a debate about whether spanking children is appropriate. Group leaders should avoid getting into arguments about this issue while at the same time emphasizing that physical abuse of children is never acceptable. Group leaders can coach parents to always avoid spanking their children when they (the parents) are angry. We encourage parents to work to find ways of disciplining their children other than spanking.

MODIFICATIONS

- This activity can be presented as either a writing activity, a creative drawing activity, or a combination. One variation is to have group members answer the questions with written answers, and then hand out another sheet and ask them to draw what they learned about anger as a child, what they want to teach their children about anger, and what they need to do to effect this transition. This variation of the activity is helpful in identifying specific steps people can take to move in the direction they want.

CULTURAL ISSUES

- Families have different norms for expressing anger, and these can be impacted by cultural background. Group leaders must be aware that some expressions of anger are acceptable in some cultures and not in others. Some families, for example, may shout more than others, and this can be acceptable in some families and not in others. Showing a genuine curiosity and interest can help draw out these differences.
- Group leaders should always emphasize the importance of a clear no-violence policy.

FOR GROUPS WITH YOUNG CHILDREN

- This activity is designed to be facilitated in groups with adults. However, once completed, group leaders can facilitate conversations between parents and children about anger in their families. If there have been inappropriate expressions of anger in the family, parents can talk about this (in group) with their children present. This can be a step toward helping families talk about difficult issues that often, up to this point, have been unspoken.
- Part of this conversation can be parents letting the children know what changes they will continue to make.
- Group leaders should encourage families to continue these conversations in family therapy sessions that hopefully are a part of the family's treatment.

WHAT I LEARNED ABOUT ANGER IN MY FAMILY AND WHAT I WANT TO TEACH MY CHILDREN

What I learned about anger in my family:

1.

2.

3.

What I want to teach my children about anger:

1.

2.

3.

What my children would say about my anger now:

1.

2.

3.

What I want my children to say about my anger:

1.

2.

3.

What I can do to make this happen:

1.

2.

3.

Strategies to Manage Anger

Activity: Strategies to Manage Anger Worksheet

Activity Mode: Psycho-education (worksheet)

RATIONALE

Anger is a common experience and concern in the families we work with. With substance abuse involved, anger can often escalate and become violent, harming everyone in the family. Encouraging a conversation among family members about how to manage anger in safe, appropriate ways is essential.

It is important for clinicians to assess for domestic violence in all couples. Sensitivity to the possibility of danger in working with couples with violence related to power and control is critical. It would be inappropriate and possibly dangerous to work with such couples in this family group. Careful assessment for domestic violence must be completed before admitting couples to the group.

That being said, we work with numerous couples who ask for help with managing their angry outbursts and acknowledge that the use of substances can often lead to escalations of conflict that lead to violence. Acknowledging and talking about this pattern can be helpful, and skills to help with this can be learned and practiced.

OBJECTIVES

- To facilitate open discussion about anger and help group members learn strategies for effectively managing anger
- To allow a safe environment for people to explore their own experience of anger and how it has impacted their lives
- To learn strategies that will help group members avoid escalation of anger
- To normalize anger as an emotion everyone feels

- To experience a relaxation exercise and gain an understanding of how relaxation can help us cope with anger
- To build hope for people who experience anger as out of their control

MATERIALS

- Strategies to Manage Anger worksheet
- Pens and pencils

OPENING QUESTIONS SUGGESTIONS

- Use two words to describe how anger is in your life.
- What is something you do to help manage your own anger?
- Has your anger changed since you (or your family members) stopped using substances? How?

METHOD

1. Ask for group members' experiences with anger in their families.

2. Pass out Strategies to Manage Anger handout. Continue to facilitate discussion on anger; review strategies to manage anger. Point out that anger is natural and can be healthy. Differentiate between anger and violence.

3. Go through each strategy on the sheet and ask group members to write down and share ideas about how they can use each strategy. Note that another week, we will spend more time practicing Active Listening.

4. Lead relaxation exercise (in Strategies handout) and facilitate discussion of the importance of learning to relax and practicing that skill.

5. Hand out What I Learned About Anger and What I Want to Teach My Children worksheet. Have group draw or write their responses, and then give time to share responses.

GROUP DISCUSSION AND PROMPTS

- Talking about times people have *not* managed anger well can be helpful, though group leaders should be cautious of shaming group members. Normalizing these experiences can be useful. If a group member describes a time they did not manage anger well, group leaders can help them identify things they can do differently next time.
- Group leaders must emphasize that violence is *never* appropriate.
- It is common for group members to say that relaxation exercises don't work for them. Group leaders should emphasize the importance of practicing these exercises when individuals are not angry. Like most things, the skill of being able to relax can be learned, but it takes practice. It can be helpful for group leaders to help individuals identify a regular time they can practice relaxation during the day.

AVOIDING PITFALLS

- As mentioned above, these strategies are not for use in families that have domestic violence.
- Sometimes group members can be unwilling to follow relaxation exercise directions. Speaking to group members directly and asking them to participate can be helpful.
- This topic can be difficult for individuals who have a background of trauma, which many of our clients have. Group leaders should acknowledge and normalize this experience and emphasize that those are issues that can be talked about in group. At the same time, we must be aware that some of our group members may be so traumatized that speaking about these issues in group may not be helpful. In such cases, it is especially important that the individual take part in individual/family therapy as part of their treatment program to help address these issues. Also, group leaders should express willingness to stay and speak individually with group members.
- It can be tempting to blame anger on substance use. Though substance use can escalate and increase anger and violence in the home, group leaders should emphasize that it is not okay to blame anger on substance use. People should be encouraged to take responsibility for their behaviors whether or not substances are used.
- Group members can sometimes minimize the impact of anger on others. Group leaders should address this pattern as it arises in group.

MODIFICATIONS

- Group can be divided into pairs to complete this worksheet. This can sometimes stimulate more discussion.
- Group leaders can plan to implement the Active Listening lesson either the week before or the week after this lesson. It can be helpful to have more time to implement this skill.
- Group leaders can integrate having group members perform skits as part of this topic, acting out inappropriate responses to anger, followed by appropriate responses. Again, special attention should be paid to ensure that group members are not retraumatized by seeing angry interactions portrayed.
- A game can be developed in which one group acts out or describes angry interactions, and another group act out a more appropriate way of handling the same situation. For example, one group could act out responding angrily to someone who rear-ended their car, and the other group could show an appropriate response to the same situation. Again, special attention should be paid to ensure that group members are not retraumatized by seeing angry interactions portrayed.

CULTURAL CONSIDERATIONS

- Group leaders must be aware that one's cultural background can have a great impact on how individuals experience and express anger. Understanding this and expressing genuine interest and curiosity in these experiences can foster healthy sharing and conversations in the group. At the same time, group leaders must be clear that physical violence is not acceptable under any circumstances.

FOR GROUPS WITH YOUNG CHILDREN

- This exercise can be very appropriate for groups with children. Managing anger is a skill children need to learn. Adults should be aware of their language, and group leaders should ensure that appropriate topics are talked about in front of children.
- The worksheet can be completed by children and adults, and both can identify skills to help manage anger.
- This can also be an appropriate venue for adults to acknowledge their angry actions and apologize to their children. Group leaders should be aware of these situations and encourage parents to continue these conversations in family therapy sessions (which should be ongoing as a part of family treatment).

STRATEGIES TO MANAGE ANGER

Read each of the following strategies and write down some ideas of ways you can implement them in your own life.

1. Look after yourself

   _____

   _____

   _____

   _____

2. Change your thinking! Instead of telling yourself, "Oh, it's awful, it's terrible, everything's ruined," tell yourself, "it's frustrating, and it's understandable that I'm upset about it, but it's not the end of the world and getting angry is not going to fix it anyhow."

   _____

   _____

   _____

   _____

3. Learn to relax and breathe (this takes practice!)

   _____

   _____

   _____

   _____

4. Listen—don't retaliate!

   _____

   _____

   _____

   _____

5. Avoid defensiveness and fighting back

_____

_____

_____

_____

6. Active listening can help!

Active listening exercise: Practice listening to your partner and repeating what s/he says until your partner says s/he thinks you heard her/him. Then switch and have your partner repeat what you say until you feel s/he has heard you. Paraphrasing is okay—but responding is not. (Active listening will be practiced more in another lesson.)

7. Use humor.

_____

_____

_____

_____

8. Change your environment—personal time—fight in the bathroom—go outside—TIME OUT!

_____

_____

_____

_____

Dangerous Thinking

- Avoid words like *never* or *always* when talking about yourself or someone else.
- Remind yourself that getting angry is not going to fix anything—it won't make you feel better (and may actually make you feel worse).
- The world is not out to get you; you're just experiencing some of the rough spots of daily life.
- Saying "I would like" something is usually more productive than saying "I demand" or "I must have" something.
- How you talk to yourself makes a difference—being aware of your self-talk can help you change how you experience situations.

Ways I can change my thinking:

_____

_____

_____

_____

Strategies that work for me:

_____

_____

_____

_____

What I most need to work on:

_____

_____

_____

_____

What is underlying the anger—fear? desire to be closer? hurt?

_____

_____

_____

_____

Relaxing Can Help (Group leaders can lead this relaxation exercise, or give copies to group members to take home to practice by themselves)

Exercise: Sit comfortably in your chair and close your eyes if that is comfortable; otherwise, pick a spot on the floor and focus gently. Breathe out completely, letting all the air out of your lungs. Breathe in slowly and fully, feeling your diaphragm and chest expand. Breathe out again, completely. Continue this slow, easy, full breathing. As you breathe out, feel stress and tension leave your body; feel your body start to relax. Allow the tension to leave with each breath. Continue this slow, easy, full breathing. Allow the tension to leave with each breath. As you breathe in, breathe in peace with each breath. Breathe out stress and tension; breathe in peace. Feel your feet heavy on the floor, your legs against the chair, feel your hands as they rest and relax. Feel the tension leave your neck and shoulders as you breathe. This feeling of being relaxed can help you when you become upset or agitated. It takes practice to remember to breathe and relax, and when we practice this when we are not upset, then it becomes easier to remember how to relax when we are upset. Now take one more minute of quiet, then begin to move your body, open your eyes, and come back to the group.

Conflict in My Family

3

Activity Title: Conflict in My Family

Activity Mode: Psycho-education (worksheet); Expressive arts (drawing)

RATIONALE

Many families we work with struggle with expressing anger in ways that are not harmful. This activity works to help families identify how they learned about dealing with conflict in their families, how those lessons impact how they express anger today, and ideas for expressing anger in more effective ways.

Group members are encouraged to consider their family of origin's ways of dealing with conflict and how that impacts how they deal with conflict in their lives today. Likewise, parents are encouraged to consider how their own ways of addressing conflict impact their children. This activity continues the idea that every family argues and that there are ways to argue that can be productive and helpful. It also continues to emphasize that violence is never acceptable and works towards helping group members acknowledge this and find ways of arguing that are not harmful.

The activity includes a worksheet as well as an expressive arts activity of drawing feelings group members experienced when, as children, they heard their parents arguing. This exercise can serve to help remind adults of the impact they have on their children.

OBJECTIVES

- To point out that conflict in families is normal and can be healthy and positive
- To strengthen a commitment to nonviolence
- To help group members learn ways of managing conflict fairly and productively
- To help group members identify healthy and unhealthy ways of managing conflict
- To help group members acknowledge impact of their parents' ways of managing conflict
- To help parents acknowledge impact of how they manage conflict on their children

MATERIALS

- Conflict in My Family worksheet
- Pens and pencils
- Crayons, markers, etc.

OPENING QUESTION SUGGESTIONS

- What was it like for you when your parents argued?
- What do you do when you are angry?

METHOD

Section 1

1. Ask group members where they learned how to argue. Say, "What was arguing like in your family?" Point out that families deal with conflict differently (in some families, yelling is okay and in some it is not, for example).

2. Emphasize that all relationships involve conflict, and learning healthy ways of dealing with conflict can be very helpful. Ask the group how they learned their own ways of dealing with conflict. Group leaders can point out that we initially learn this from our families, and sometimes what we learn is not very healthy. Emphasize that it is possible to relearn these skills.

3. Hand out Conflict in My Family worksheet and allow time for the group to work on the first three questions. Encourage group members to work with their family members (or with another group member) to stimulate conversation about each question.

4. Encourage group members to share answers to these first three questions. Try to allow each person in the group to respond in some way.

Section 2

1. Facilitate expressive arts exercise. Encourage group members to sit quietly, with eyes closed if they choose. Let the group know that this exercise is about remembering an experience from the past, and that it can be painful for some people. Encourage people to look after themselves and go as far with the exercise as they are comfortable. Emphasize the importance of self-care and that if the exercise feels too difficult, talk to a group leader.

2. Group leaders should read out loud (slowly) the instructions: Take a moment and remember a time when you heard your parents (or other adults) arguing. Remember how you felt as you heard them. Imagine where you were, what you were doing. Remember what you felt.

3. When you have a memory of your feelings, imagine how those feelings would look if they were an image. What color would they be? What shape? Take some time to draw that image on the other side of this page. If no image came to your mind, that's fine—just start drawing.

4. Emphasize there is no right or wrong. The drawing does not have to look pretty and will not be judged. The point is for the group to recognize their feelings.

5. Ask the group to come back together, and give an opportunity for group members to share their drawings and their experiences with that exercise. Ask if anyone learned anything they didn't know. Ask if anyone had strong feelings during the exercise. Emphasize that if strong feelings arise, group leaders are available to help process those feelings. Let group members know they can share as much or as little as they want. If someone is not willing to share his or her drawing at all, group leaders can ask what the experience was like for him/her.

GROUP DISCUSSION AND PROMPTS

- Emphasize that many of us do not learn healthy ways of resolving conflict as we grow up. As a result, we often do not know how to resolve conflict in ways that are productive. This can be learned!
- Group leaders should continue to emphasize that violence is never acceptable.
- Encouraging group members to talk about their experiences as children can help group members acknowledge the impact of their childhood experiences on their lives. This can also be helpful in helping parents recognize the impact of their own behaviors on their children.
- Often, people who were initially hesitant about participating in the expressive arts activity find it can be an enjoyable process and can result in new insight. Asking group members if anyone had this experience can be useful. Normalizing nervousness about drawing can help alleviate some of the discomfort. If group leaders share their own experiences of shyness with expressive arts activities, the group often responds positively.

AVOIDING PITFALLS

- It is not uncommon for group members to talk about spanking children during this conversation. Though there are different thoughts about spanking children (the authors do not endorse spanking children), there is no question that spankings should never be given when parents are angry.
- There can be a tendency for group members to fall into telling "war stories" of their experiences with their parents fighting. Group leaders should guide the conversation away from this and toward acknowledging the difficult situations children can be in.
- Group members often have trauma from childhood experiences, as well as guilt from exposing their own children to harmful situations. Group leaders must be aware of these responses and let group members know they are willing to stay after group to address strong emotions that can arise. These emotions can arise strongly during the expressive arts piece of the activity.
- Group members will often balk at expressive arts activities. Emphasizing that artistic talent is not necessary can be helpful. Letting group members know that they will not be required to share their drawings can also be helpful.
- Asking the group to complete the activity in silence can add to the power of the exercise and also avoid group members encouraging each other's negative thoughts about the exercise.

MODIFICATIONS

- The expressive arts piece of this exercise can be expanded to include a drawing of group members when they are angry themselves. There could be two phases of the activity: the first as written above (draw how you felt when your parents were angry), and the second with the instruction to draw how you feel when you are angry.
- This activity can be separated into two weeks, with one week used to work on the psycho-educational piece and the second week for working on the expressive arts piece.

CULTURAL CONSIDERATIONS

- Group leaders can encourage conversation about how anger is expressed in different cultures.
- Group leaders should emphasize that violence is never acceptable.

FOR GROUPS WITH YOUNG CHILDREN

- Talking about anger can be very important for children. Normalizing anger as an emotion that everyone has can be helpful, as well as emphasizing that we all need to find ways of managing our anger in ways that are not hurtful—to us or to others.
- Group leaders must be sensitive to the fact that many families in the group may have experienced physical violence. Ensuring safety in the group is essential, as well as helping parents let their children know that it is safe to talk about these issues. It could be helpful to have conversations with the parents separately if the group leaders believe that talking about these issues could be difficult for the family. Children's safety must be an absolute priority.
- Children are often less likely than adults to balk at the expressive arts piece of this exercise.

CONFLICT IN MY FAMILY

How did your parents argue (loudly, privately, physically, etc.)?

_____

_____

_____

_____

What did this teach you about arguing and conflict?

_____

_____

_____

_____

How do you argue in your current relationships? When you have an argument with someone, is it usually productive?

_____

_____

_____

_____

For parents: How is the way you argue impacting your children? What are they learning? What would you like them to learn differently?

_____

_____

_____

_____

Take a moment and remember a time when you heard your parents arguing.

Remember where you were and what was going on. Remember how you felt.

On the other side of this page, take some time to draw these feelings.

Example: How I Felt When My Parents Argued

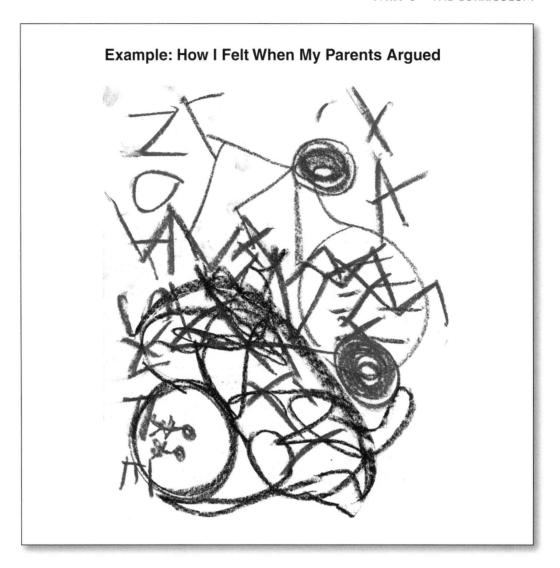

Making Arguing Work

4

*Activity Title: Fair Fighting
Rules Worksheet*

*Activity Mode: Psycho-
education (worksheet)*

RATIONALE

This topic begins to address the idea that there are healthy and unhealthy ways to argue. We normalize the fact that all families have disagreements, and they argue. Most group members will recognize that sometimes arguments can be productive and sometimes not, and most respond to the idea that it may be possible to learn to argue in ways that are less harmful and more productive. This activity helps group members begin to consider these ideas, as well as teaches a concrete tool for talking about difficult subjects (when you ____, I feel _____. I wish that you would _____.).

Safety is framed as being essential, and the group is encouraged to consider when a time out would be necessary to ensure safety.

Group members are then asked to develop a list of fair fighting rules for their own families.

OBJECTIVES

- To point out that arguing is normal, and can be healthy and positive
- To help group members learn ways of arguing fairly and productively and identify particular rules that will be helpful for their family
- To help group members learn a concrete tool to help them address difficult topics successfully
- To emphasize that violence is never acceptable

MATERIALS

- Fair Fighting Rules worksheet
- Pens and pencils

OPENING QUESTION SUGGESTIONS

- What was it like for you when your parents argued?
- How do you argue? What do you do when you are angry?

METHOD

1. Ask for group members' experiences with arguing in their families.

2. Hand out Fair Fighting handout and pens and facilitate discussion. Emphasize that arguing in relationships is normal and can be healthy if managed appropriately.

3. Group leader reads each point and asks for responses from the group.

4. Break up into pairs (family members can be together). Take time for them to practice the suggestions (such as #5, "When you _____, I feel _____. I wish that you would _____.).

5. Group leaders can ask for volunteers from the group to act out examples from the worksheet.

6. Give the group time to complete their own list of fair fighting rules. Group members can work together in pairs or with family members.

7. Ask the group to share the rules they would like to have in their homes. Ask them to share if there are particular reasons they chose those rules.

GROUP DISCUSSION AND PROMPTS

- Talk about what to do if an argument is escalating: TIME OUT.

    a. When is a time out necessary?

    b. How long a time out is necessary?

    c. How do you know when to try to talk about the issue again? (Maybe wait until your therapist is available?)

    d. SAFETY IS ESSENTIAL
- In most groups, the fact that many individuals have experienced violence growing up and/ or in current relationships will arise. It is essential that group leaders consistently emphasize that violence is never acceptable and that alternatives to violence must be found. Group leaders can also give information about the domestic violence services in the area.
- Many individuals that grew up in violent homes believe that violence is normal. It can be helpful for group leaders to point this out to the group. As group members recognize they are not alone in their experiences, they are often more willing to share their beliefs and experiences.

AVOIDING PITFALLS

- Group leaders must emphasize safety and at the same time create a space that encourages conversations about beliefs. Group members that were raised in a culture of violence could come to the group believing that violence is acceptable. It would be helpful to have an open conversation about this. Group leaders must balance creating an open environment with emphasizing that violence is not acceptable in any situation.

- This topic can trigger trauma in some group members. Group leaders should acknowledge the difficulty children have being raised in violent homes, normalizing this trauma. An opportunity to speak with group leaders after group can be given. These issues should also be addressed in individual and family sessions.
- Group members may also experience shame and guilt about exposing their own children to violence during the course of their addiction. Group leaders should normalize this experience and offer hope for change. (The fact that group members are present and addressing these issues gives hope.)

MODIFICATIONS

- Group leaders could divide the group into pairs or small groups and assign each group one of the points on the worksheet. Ask them to develop two skits, one that shows the point less effectively and one that shows it more effectively. For example, one skit shows an individual talking about many issues at a time and another shows the individual speaking more effectively about one issue at a time (see worksheet).
- Group members can create their fair fighting rules list on a separate sheet of paper that can be made to be posted in their homes. Encourage them to be creative, and provide markers, crayons, and/or magazines to create a collage of their fair fighting rules. This can be especially helpful for groups with children.

CULTURAL CONSIDERATIONS

- Group leaders must be aware if there are group members that do not read well and adjust the group accordingly.
- Talking about anger and arguing in different cultures can be helpful. This can be challenging if group members try to say that physical violence is acceptable in some cultures. Group leaders must be able to acknowledge and respect cultural differences and at the same time hold a clear "no violence" policy.

FOR GROUPS WITH YOUNG CHILDREN

- This activity can be effectively implemented with groups with young children. Taking each of the concepts and acting them out is a way that children can actively participate in this idea.
- Children that have been witness to parents' violence may be afraid to talk about these issues. Group leaders must tread gently, and allow children the freedom to go as far as they want to go with these concepts. Safety must be the first priority.
- The idea of learning how to argue effectively can be very positive for children. Asking how children can use these skills in their own lives (with their friends, for example) can be useful. Children also can identify other ways of arguing that can be effective. Group leaders should model interest and curiosity in children's thoughts, experiences, and ideas.
- Many children experience time outs as a consequence from their parents. Soliciting children's thoughts on whether this is a helpful intervention can be extremely interesting. Asking children about what helps them calm down when they are upset can trigger interesting conversation and sometimes profound insights.

FAIR FIGHTING RULES

Some Things to Try

- Focus on solving a problem, not winning (what is the difference?).
- Address one issue at a time.
- Stay focused on the present (if you need to talk about past issues, set a time to do that).
- Avoid globalizing ("you never...," "you always...").
- If things are too volatile, have a neutral party present.
- State the problem clearly.
- Avoid blaming the other person—use "I" statements.
- When you _____, I feel _____. I wish that you would _____.
- Be willing to listen to what the other person has to say.
- Summarize what you hear the other person saying. That means you must really listen, not just think of what you will say next (active listening)!
- Take turns speaking and listening.
- Don't interrupt, talk over, or make comments while the other person is speaking.
- Watch your nonverbal expressions, too (rolling eyes, smirking, yawning, etc.).
- Focus on the problem, not the person.
- Time out—make an agreement about when it is okay to take one, as well as an agreement of how and when to re-engage.
- Avoid making threats.
- Other:

_____

_____

_____

_____

Time Out

When: _____

How long: _____

When do we try again? _____

OUR FAMILY'S FAIR FIGHTING RULES

What are some rules you would like to make for fair fighting in your home? Make sure you include what is *not* allowed (no hitting, for example). Note that there will be different rules for different homes, and some of this can be cultural. Some families, for example, are very comfortable with shouting, whereas others are not.

Our Family's Fair Fighting Rules

1. _____
2. _____
3. _____
4. _____
5. _____
6. _____
7. _____
8. _____
9. _____
10. _____

Not Allowed in Our Family

1. _____
2. _____
3. _____
4. _____
5. _____

Curriculum Section VI:
Communication

Working Together

1

Activity Title: Words and Story

Activity Mode: Expressive Arts (writing)

RATIONALE

This activity involves working in pairs to create a story about group members' families and their recovery. Initially, group members will have a discussion about what it is like to work with other people. Many of our group members are isolated and have limited experience working with others. This activity is designed to help bring back the experience that working with others can be satisfying.

In our experience, the expressive arts piece of this activity can result in group members developing stories that are creative and powerful. In each group for which we have facilitated this exercise (despite group members' initial negative responses to hearing about the exercise), this has been the case. Expressive arts exercises often help individuals access experiences and truths that are not available on an intellectual level and can bring to light profound insights and experiences. This can be particularly powerful when families work together on this project.

OBJECTIVES

- To introduce idea of the importance of working together
- To encourage group members to think about how their families work together
- To give an opportunity for group members to work together
- To help people think about how they work best with others
- To encourage openness to the possibility of changing how we work with others
- To help group members develop insight into their experience of family
- To help group members develop insight into their experience of recovery

MATERIALS

- Blank paper
- Pens, pencils
- Working Together worksheet

METHOD

1. Ask individuals in the group how they work with others (is it easy? challenging? are they leaders? how do they deal with friction? etc.).

2. Encourage group to share experiences when working well with others was helpful and when not working well with others made things difficult.

3. Divide group into pairs. Encourage families to stay together (can be more than two in a group).

4. Hand out Working Together worksheet. Ask each pair to make a list of 16 nouns (people, places, or things). One idea is to alternate, with first one person picking a word, then the other. Note that the words must be appropriate (no swear words). Instruct that there be no talking during this part of the exercise—just write the words on the paper. (There is a space to write the words on the worksheet.)

5. Then instruct the teams to make a story that involves their families and recovery, using each of the words they wrote down (both lists). It is important not to give this part of the instruction until after they have completed the word-list section.

6. Invite participants to share their stories.

   a. As they finish writing the stories, encourage them to think about and respond to the questions on the sheet.

   b. Ask for volunteers to read their stories (not a requirement to share, but have each group member say something about the experience).

   c. Ask what it was like to do this activity, to be the leader and the follower, and to write the story. Did they trust their partners?

   d. Ask what role people took in the process (Were they leaders? Who did the writing? Did one person tend to take charge? etc.).

   e. What happens when people work together? Is it more fun? less fun? easier? harder?

AVOIDING PITFALLS

- Some group members may be confused by the instruction to develop a list of nouns. Describing the words as "person, place, or thing" can be helpful. If this is confusing, this limitation on the words can be removed, and the list can be of any words that come to mind.
- If making the list is difficult, group leaders can use word association to help group members develop the words on their lists. Group leaders can say words and have the group write down what comes to mind. However the list is created is fine.
- As with other exercises, group leaders must be aware and respectful of group members who do not read and write. One way we manage this is for the group leaders to offer to work with those individuals, or have them pair with an appropriate group member. It can be helpful to have a conversation privately with the group member to ask how the group leaders can help with this situation. It also would be possible for less literate clients to write an outline and tell the story verbally.

MODIFICATIONS

- One modification to this exercise can be to have groups make up the lists together, then write their stories individually. The groups can then get back together to share their stories with each other. This can be an interesting experience of how individuals starting with the same information (words) will create entirely different stories.

CULTURAL CONSIDERATIONS

- There are no specific cultural considerations for this exercise. However, group leaders must always be aware of and respectful of different cultural experiences. Many of the stories can reflect different cultures, and group leaders should reflect with interest and curiosity.
- Different family cultures will be expressed in this exercise. Individuals growing up in the same family often have very different experiences. This exercise can serve to further awareness of these experiences for families.

FOR GROUPS WITH YOUNG CHILDREN

- This exercise can be very appropriate for groups with young children. The stories can be written by the parents, with children dictating. Or children can do the actual writing. Another option would be for the parents to work on one story and the children on another and then both share them. Children can often give powerful insights on how their family works together.

WORKING TOGETHER

_____

_____

_____

_____

_____

_____

_____

_____

_____

_____

_____

_____

_____

_____

_____

_____

_____

_____

_____

_____

_____

1. When doing the activity, were you the leader or the follower? Why?

   _____

   _____

   _____

2. What role did you take? Who did the writing? Did one person tend to take charge?

   _____

   _____

   _____

3. What happens when people work together? Is it more fun? less fun? easier? harder? What makes it easier or harder for you to work with others?

   _____

   _____

   _____

4. Do you tend to be easy or difficult to work with? Do you prefer working alone or with others?

   _____

   _____

   _____

WORDS

_____   _____   _____   _____

_____   _____   _____   _____

_____   _____   _____   _____

_____   _____   _____   _____

2 Assertive Communication

Activity Title: How Assertive Am I?

Assertiveness/Aggressiveness Scenarios; How to Ask for What I Want

Activity Mode: Psycho-education (worksheet); Expressive arts (acting); Role Plays

RATIONALE

Often, our group members have limited social skills that help them get what they want. Many grew up and lived their lives seeing people around them act aggressively much of the time. The idea of assertive communication can be a new concept for many.

The importance of this exercise is that it can help group members recognize a difference between assertiveness and aggression and, hopefully, begin to increase assertiveness and decrease aggression in their lives. This is particularly important for families with children, since children learn what they see.

The activity incorporates an assertive self-assessment that can help group members identify ways they can move toward being more assertive in their lives. The second section of the activity involves acting out skits in which group members portray both assertive and aggressive responses to situations. This activity helps make the concepts concrete and can help group members identify and practice assertiveness skills.

OBJECTIVES

- To help group members distinguish between assertive communication and aggressive communication
- To help group members increase assertive communication and decrease aggressive communication
- To identify communication patterns in families, which are passed from one generation to the next

- To help group members develop new patterns of communication that will then be modeled for their children
- To give group members an opportunity to practice assertive communication

MATERIALS

- Handout: Assertiveness Self-Assessment
- Assertive scenarios
- How to Ask for What I Want worksheet
- Pens, pencils

OPENING QUESTION SUGGESTIONS

- Is it difficult to ask for what you want? When is it difficult and when is it not difficult?
- How do you ask for what you want (directly, indirectly, etc.)?

METHOD

Part 1

1. Facilitate discussion of the differences between assertiveness and aggression. With the group's input, list differences on the board in two columns. (See Assertive Self-Assessment for ideas to help the group get started.)

2. Encourage group to give examples from their own lives, both currently and experiences as children.

3. Ask about difficulties being assertive and why people are sometimes aggressive.

4. Allow time for group members to complete self-assessment, including listing three actions they can take to become more assertive.

5. Gather back into the group and facilitate discussion about how group members completed the assessment. What questions stood out? What did they realize as they answered the questions? What did you learn from this?

6. Ask group members to share one thing they can do to become more assertive.

Part 2

1. Break group into small groups and divide scenarios. Have each group act out each scenario twice, once using assertive skills and once aggressively. Differences should be noted in the body language, words, and tone of voice, as well as the results of the interaction.

2. After about 5 minutes, ask the small groups to act out their skits in front of the group. Ask for feedback from the group. Also ask the small groups how it felt when they were acting aggressively. Assertively? What did the other small-group members feel with each?

Part 3: How to Ask for What I Want

1. Divide the group into pairs, have each person identify something they can ask for, and use the format "When you _____, I feel _____. Could you please _____?" They can use situations from their own lives or a hypothetical situation.

2. Regroup and have some of the pairs model asking in this format for the group. Ask how it felt to ask for something that way (awkward, possibly). Point out the different responses they get when they ask this way as opposed to demanding.

GROUP DISCUSSION AND PROMPTS

- Group leaders must recognize that some of these concepts can be new for group members. Acknowledging this can be helpful, and encouraging more conversation about the impact of growing up in homes in which aggressive communication (including violence) was modeled impacts children immensely. Further communication about changing these behaviors because of the impact on children can be helpful.
- Group members can carry a weight of guilt for exposing their children to hurtful situations. Acknowledging and normalizing this guilt and acknowledging group members' willingness to change now can be helpful. Other group members can often offer encouragement by sharing their experiences of making these changes. Group leaders should be sure that family therapy is recommended for families in these situations.
- It can be interesting to ask the women in the group about their experiences with assertiveness. Culturally, it is not uncommon for a strong, assertive woman to be labeled a "bitch." A conversation about this among group members can be useful.

AVOIDING PITFALLS

- Group leaders should be aware that the skits, when acted out aggressively, remain within limits. Many of our group members come to us with a background of trauma, and we must be cautious that they are not retraumatized in group. Limits must be set on the aggressive actions during the skits (no touching, limited shouting, etc.). Rather than act out these extreme behaviors, the small groups can say what they would do (for example, "I would hit him") or play-act (swing but not make contact).
- As mentioned above, group leaders should be sensitive to parents in the group that feel guilt because of their behaviors as parents and respond to this in a supportive way.

MODIFICATIONS

- All three parts of this exercise may not be possible to complete in one group. If this is the case, this activity can be stretched out over two weeks. This can be a helpful option, since spending more time on Part 3 and having group members practice asking for what they want can be productive.
- The assertiveness scale can be implemented as a physical activity rather than a written one. With the group in the middle of the room, group leaders read the statement and

ask everyone for whom the statement is true to go to one end of the room and everyone for whom it is false to go to the other. The group leaders can then ask what made group members choose which end of the room they went to.

CULTURAL CONSIDERATIONS

- There can often be conversations that arise about what is and is not okay based on group members' cultural backgrounds. We take a stance of openness and curiosity and ask group members to share more about their cultural background and its impact on these issues. We are also always clear that violence is never acceptable.
- Women's issues can arise during this topic—and if they don't, we generally bring them up. It is important to us that the different expectations our society has for women be talked about in group. Women generally are eager to share those experiences.

FOR GROUPS WITH YOUNG CHILDREN

- This can be an excellent exercise for families with children to work on together. Changing the patterns in families is always a goal for us, and this particular exercise can help children identify harmful ways of communicating that may be very comfortable for them. Asking for what they want is a skill many of the children in our families have no experience with, and this activity can help build those skills.
- This can also be a good opportunity for parents to acknowledge past behaviors that have been harmful and to let their children know it is okay to talk about that and that they are working on changing those patterns.
- It can also be very helpful for individual family therapy to be available for these families that are working on such powerful changes.

HOW ASSERTIVE AM I?

Thinking about the following situations can help you assess where you are in terms of being assertive. For each statement, mark where you are on the scale of 0–10:

I can ask for help without feeling guilty or anxious.

0	1	2	3	4	5	6	7	8	9	10
Never										Always

I can say no easily.

0	1	2	3	4	5	6	7	8	9	10
Never										Always

I confidently express my honest opinions.

0	1	2	3	4	5	6	7	8	9	10
Never										Always

It is easy for me to express my feelings, even when I am feeling strong emotions.

0	1	2	3	4	5	6	7	8	9	10
Never										Always

When I express anger, I do so without blaming others for making me mad.

0	1	2	3	4	5	6	7	8	9	10
Never										Always

I speak up, even when I am in a group.

0	1	2	3	4	5	6	7	8	9	10
Never										Always

I admit my mistakes.

0	1	2	3	4	5	6	7	8	9	10
Never										Always

I tell others when I am not comfortable with their behavior.

0	1	2	3	4	5	6	7	8	9	10
Never										Always

I am comfortable meeting new people.

0	1	2	3	4	5	6	7	8	9	10

Never   Always

I can talk about my beliefs without putting down other people's beliefs.

0	1	2	3	4	5	6	7	8	9	10

Never   Always

I am confident that I can learn something new.

0	1	2	3	4	5	6	7	8	9	10

Never   Always

I believe my needs are as important as the needs of others and I am entitled to have my needs satisfied.

0	1	2	3	4	5	6	7	8	9	10

Never   Always

Taking the above information into account, what are three things I can do to help move toward being more assertive in my life?

_____

_____

_____

_____

_____

_____

ASSERTIVENESS/AGGRESSIVENESS SCENARIOS

Scenario 1

A coworker has been criticizing how you do your work (behind your back). You want to discuss this with that person.

Scenario 2

You need to ask a colleague to be quiet. His/her constant chatter and loud voice are disrupting your own work.

Scenario 3

You are in the parking lot in an expensive car you have borrowed from a friend for the day. As you are reversing the car out of its space, another car runs into it.

Scenario 4

As you are backing out of your parking space, a woman thumps the roof of your car and shouts, "Stop!" She then tells you that you almost knocked her 5-year-old child down.

Scenario 5

A friend of yours has told someone else a confidence that you asked him/her not to tell.

HOW TO ASK FOR WHAT I WANT

The following phrasing will help you ask for what you want in a way that is likely to help you get it! Notice the request is clear, and the format avoids blaming the other person. Try it!

When you _____,

I feel _____.

Could you please _____?

What We Haven't Said

3

Activity Title: Letter to Family Member

Activity Mode: Expressive Arts (writing)

RATIONALERATIONALE

Most of our clients come to us after years of living in or with active addiction, which resulted in behaviors that hurt families. Often, our clients have lied to, stolen from, and betrayed their loved ones. Thus, clients often have many regrets and a great degree of shame about how they have treated their families and others that have cared about them in their lives. Part of being in recovery is acknowledging this and eventually taking some steps to make amends for these behaviors.

This activity gives group members a chance to identify something they would like to say to a family member or friend that they have not said before. Though we do not restrict or set a rigid rule as to what they should write about, most clients choose to either apologize for something they have done or acknowledge and thank someone for the help they have given.

This activity can trigger strong emotions. As in all activities, clients should be encouraged to share what they are comfortable sharing. Normalizing feelings of sadness and shame surrounding these issues can be invaluable. It can also be helpful for group members (and family members) to hear other people's stories and recognize that they are not alone with their own.

OBJECTIVESOBJECTIVES

- To create an opportunity for group and family members to share something important that they have not shared before
- To encourage recognition of the importance of communication and talking about difficult subjects as ways of feeling better yourself rather than ways of changing the other person
- To normalize and acknowledge difficulty in doing so
- To normalize the fact that most addicts have damaged relationships, particularly family relationships
- To create a safe environment for expressing emotions
- To create an experience in which sharing difficult emotions can have positive results

187

MATERIALS

- Paper and/or "I Need to Tell You Something" exercise
- Pens, pencils

OPENING QUESTION SUGGESTIONS

- Who is someone you have apologized to in your life? Was it easy or difficult?
- What is the value of an apology?
- What is an example of something you can change and something you cannot change?

METHOD

1. Normalize the pain addicts and families experience as a result of addiction—this is something that all families experience and have difficulty with.

2. Normalize doing actions that hurt others (particularly family members) while in active addiction.

3. Normalize difficulty talking about these painful experiences.

4. Let the group know that as recovery progresses, it can be helpful to speak about some of these things. Often, both the addict and the family members have things they want to say but are afraid to. Point out that the purpose of talking about these issues is not necessarily to change the other person's feelings or actions but rather to help yourself. The Serenity Prayer ("God, grant me the serenity to accept the things I cannot change, the courage to change the things I can, and the wisdom to know the difference") can be useful in helping individuals differentiate between what can be controlled (what you do) and what cannot be controlled (what others do).

5. Present writing a letter as a possibility of a safer way of saying difficult things (rather than saying them in person).

6. Let the group know that tonight, they will write a letter to a family member to express something they have not shared before. Let them know they have complete freedom about what they want to write about; there is no requirement it has to be anything major, just something new.

7. Mention that this can be a powerful experience and can bring up strong emotions. Encourage them to ask for help if anyone is feeling overwhelmed.

8. Let them know that after group, they will have an opportunity to share their letters with the group, but they will not be required to. They will be asked who they wrote their letter to and the topic, but they only have to share what they are comfortable with.

9. Allow about 15 minutes to write the letters. Ask for silence in the room during this process.

10. Bring the group back to the circle, and ask each person to share his or her letter and/or experience of writing.

11. After the group shares, again normalize the experience of strong emotions from such an exercise, and give an opportunity to talk about that. Also let the group know that if anyone needs to talk after the meeting, they are welcome.

GROUP DISCUSSION AND PROMPTS

- It can be helpful to emphasize that this letter can be about anything. Many people decide to write an apology letter, but any other topic is just acceptable as long as it is in some way meaningful to you.
- It can also be helpful to emphasize to group members that it is not necessary to give the letter to the person; that decision is completely up to the writer. An important part of this exercise is the writing itself, and not delivering the letter does not take away from that process. In some cases, it might be helpful for the writer to burn the letter (for example, if the letter was written to an abuser) as part of a ritual of healing and letting go. Group leaders can work with individual group members with their situations.
- Ask the group about their experience with this exercise. Was it easy? Difficult? Did strong emotions come up? Was it moving to hear other group members' experiences?
- Encourage the group to talk about the goal of writing the letter, bringing up the idea that the goal is not to change anyone else's behavior. For example, an apology will not necessarily result in forgiveness. On the other hand, writing a genuine apology may help the group member with her or his process of healing.

AVOIDING PITFALLS

- Since this is an exercise that requires writing, it is important to be aware of any group members that cannot write and to give them extra help and support without making them feel uncomfortable. This can be a challenge. One suggestion is to have a group leader work with this client and write the letter for him/her. It is also sometimes helpful to give the option of drawing rather than writing words.
- Our experience is that writing the letters in silence can encourage focus on the exercise and help avoid side conversations that can take away from the power of this exercise.
- This exercise can trigger strong emotions. Group leaders must be aware of this and create an environment in which group members can easily ask for extra support. Leaving adequate time for processing at the end of this exercise is important.
- Group members may describe situations in which they have apologized and not been forgiven. This experience can open the door to a productive discussion of what we have control of, what we don't have control of, and the reasons for doing this activity (which are about group members' own healing processes rather than trying to change anyone else's behaviors).

MODIFICATIONS

- This exercise can be presented as a drawing exercise rather than a writing exercise, in which group members are asked to draw something they want to express to someone in their lives. Group members would then be asked to share their drawings as they are comfortable, or at least share what their drawings were about.

CULTURAL CONSIDERATIONS

- As discussed above, group leaders must be aware of group members that do not read or write and provide support and help with this activity. The modification (above) of using drawing rather than writing can be helpful in this situation.
- Cultural norms surrounding what is shared and what is private vary, of course. Apologies, for example, are experienced differently in different cultures. Group leaders must be sensitive to these differences and can encourage group members to talk about their own experiences by expressing genuine interest in and respect for clients' cultures. In the mountain culture in which we live, it is common for group members to be very private, and sharing personal family experiences and emotions can be difficult. Our experience is that acknowledging and respecting these different experiences and values can be very helpful in building safety in the group, often resulting in a willingness to share.

FOR GROUPS WITH YOUNG CHILDREN

- This activity can be helpful in groups with children. Group leaders can ask children to write a letter to parents and parents to write a letter to children. Then they can take turns reading the letters to each other, either privately or in front of the group. This part of the exercise can be very powerful and emotional. Neither parents or children should be forced to share what they are not comfortable sharing.
- Children and parents can be coached to listen to the letters, at first without responding, and then responding with an acknowledgement of the other's willingness to share ("thank you for telling me that").
- Group leaders can set some guidelines as what to write (such as "What is something you want to say thank you for that you haven't?" or "What is something you want to say you are sorry for?" or "What is something you are angry about?"). This activity can be framed as a way to start a conversation that can be continued rather than a chance to say everything that has not been said.
- Parents should be cautioned that what their children say may be painful. Therapists should be aware of emotional reactions and willing to offer support. Family therapy can serve to help families continue to process the issues that arise.
- If the letters are shared privately, the group should then circle up, and each family has an opportunity to talk about its experience. What was it like to write the letter? To share it? What is the next step?

Date: _____

Dear _____,

I need to tell you something. _____

_____

_____

_____

_____

_____

_____

_____

_____

_____

_____

_____

_____

_____

_____

_____

_____

_____

_____

_____

_____

_____

_____

_____

_____

_____

4

Our Different Experiences

Activity Title: Answer by Moving

Activity Mode: Experiential

RATIONALE

In our experience, many group members come to group with reluctance, often thinking they are different than everyone else. After participating in group for some time, most group members move past this initial experience. As connection builds among group members, they will often share their initial experience with newcomers who enter the group.

This activity helps group members recognize they have surprising things in common with other people (including group leaders) in the group. The exercise facilitates the group answering various questions by moving to one end of the room or another so that the answers are very visible. For example, one question may be, "Did you grow up in a home where there was extreme alcohol or drug use?" As most of the group moves to the "yes" side of the room, a strong experience of that connection with others in the group is created.

The questions in this exercise generally move from mundane ("Who was born here?") to deeper questions, such as, "Who has had a close friend or family member die because of this disease?" The questions suggested here are only suggestions, and group leaders should feel free to add their own questions based on their knowledge of their own group.

OBJECTIVES

- To build connection among group members
- To allow members to share difficult experiences in a safe manner
- To offer an opportunity for group members to talk about shared difficult experiences
- To help group members see that they are not alone in their experiences
- To help group members acknowledge and develop compassion for experiences of family members

MATERIALS

- Worksheet: Questions for Answer by Moving
- A ball (8-inch diameter or larger is best)

OPENING QUESTION SUGGESTIONS

- Who is one person in this room you feel a connection with, and why?
- When you first started coming to group, what helped you feel comfortable?
- What is one thing you can do to help new group members feel more comfortable?

METHOD

1. Have group stand at one end of the room and spread out as much as possible (group will be walking from one end of the room to the other). Group leaders stand at the side of the room, about halfway. (This activity works well outside if there is a field available.)

2. Group leader explains that he/she will read a question, and let the group know which end of the room is the place to go for which answer. For example: "Which do you like, Coke or Pepsi? Coke lovers stay where you are, Pepsi lovers go to the other side of the room."

3. Once group has moved, group leaders throws the ball to any group member and asks if they would like to talk about their choice.

4. Group member then throws the ball to another group member, and this continues until all those who want to speak have done so. The ball is then passed back to the group leader, who asks the next question.

5. Group can also be asked to contribute questions for the activity.

6. Usually, willingness to share increases as this activity moves along.

GROUP DISCUSSION AND PROMPTS

- Ask the group if they are surprised at all they have in common with each other. What does that mean to them? Does it make them look at anything differently?
- Ask what it was like to participate in this activity. Did you feel emotional at any time? What was that about?
- Ask the group if anyone wants to contribute any questions.
- It can be helpful to have silence during the movement part of this activity (as group members choose their answer and move).

AVOIDING PITFALLS

- This is an activity that can become scattered if group members do not keep focused. When we go outside, often group members want to smoke, and we generally set limits on this (they would not be smoking if they were inside).

- Asking for silence during the movement piece of the activity can curb side conversations and can allow emotions to be felt during the activity.
- If group members are asking questions, group leaders must be aware of any inappropriate questions and not allow them. Questions about very personal issues (how many people were sexually abused, for example) must not be allowed.
- This activity can bring up strong emotions as those present acknowledge some of the difficult experiences they have had related to addiction, recovery, and their families. At the end of the activity, it is important to give time for processing the experiences and giving support to anyone who is triggered or feels unsafe in any way. This also highlights the need for a balance of questions that are humorous and/or "light."

MODIFICATIONS

- Once the group leaders have asked a few questions, that part of the activity can be turned over to group members. Each group member can have a chance to ask a question of the group and direct which side of the room will be for which answer.

CULTURAL CONSIDERATIONS

- There are no specific cultural considerations for this activity, though as with any activity, culturally related issues may arise and must be handled appropriately—with respect and affirmation of cultural differences.

FOR GROUPS WITH YOUNG CHILDREN

- This is an excellent exercise for groups with young children, though the questions may be different for such groups. Children would enjoy participating in making up the questions and directing the adults. This activity can be a way for children to continue to recognize that they are not alone with their experiences.

ANSWER BY MOVING

Questions for the Group

1. Who was born here (in this city, county, or state)?

2. Who was born outside of the USA?

3. Who is an only child?

4. Who has children?

5. Who grew up in a family with too much alcohol and drug use?

6. Pepsi or Coke?

7. Likes vegetables?

8. Who plays an instrument?

9. Who lost a close relative to this disease?

10. Who doesn't know anyone that died because of this disease?

11. Who grew up in a two-parent household?

12. Who lived with violence growing up?

13. Dogs or cats?

14. Who has been to inpatient treatment?

15. Who has been to jail as a result of this disease?

16. Who works with their hands?

17. Who played sports as a child?

18. Who has important things they want to say to their family?

 a. Who is willing to say some or all of those things?

19. Baseball or football?

20. Ford or Chevy?

21. Who was raised by someone other than your parents?

22. Who was raised in a home with domestic violence?

23. Who has been afraid that a loved one would die because of drugs or alcohol?

24. Who dreaded holidays as a child?

25. Who walked on eggshells as a child or an adult?

26. Who believed at some point in your life that you were responsible for another person's using or drinking?

27. Who has blamed another person for your own using or drinking?

28. Fall or spring?

29. Who is considering making amends to someone else?

30. Who has made some amends to someone else?

31. Whose life is better today than it was 6 months ago?

5

Do You Hear Me?

Activity Title: Active Listening

Activity Mode: Psycho-educational (experiential)

RATIONALE

This activity helps develop the skill of listening without immediately responding, as well as the experience of feeling "heard." This can be a challenging process for everyone, our clients included.

In this exercise, we differentiate between reacting and responding and facilitate an active listening exercise that allows group members the experience of hearing each other and feeling heard. Many of the people we work with have little or no experience of feeling heard, and often we hear people respond in positive ways to this experience. Also, helping clients listen and not respond immediately can result in a new experience of more fully understanding another person's experience.

We teach this as a skill that can be used when difficult interpersonal issues arise. Most group members can identify people or issues in their lives with which active listening can be useful.

OBJECTIVES

- To normalize having difficulty listening without reacting
- To differentiate between reacting and responding
- To help group members have the experience of actively listening to another person
- To help group members have the experience of feeling "heard"
- To help group members develop a skill to use in difficult personal situations

MATERIALS

- Active Listening worksheet
- Pens

OPENING QUESTION SUGGESTIONS

- What is an example of something that is hard for you to talk about?
- Is there someone in your life that you feel really listens and hears you? Who is that person, and what makes you feel that way?

METHOD

1. Ask the group what the difference is between reacting and responding. Ask the group to give examples of each.
   - Group leaders can point out that reacting is often an immediate, emotional response. Often, reacting can get us in trouble. Responding, on the other hand, is a more considered approach to a problem—and often results in a more positive result. Reacting harshly to a probation officer, for example, will likely result in a negative consequence, whereas responding appropriately will more likely result in avoiding a negative consequence. Group members will be able to give examples of both reacting and responding to situations.

2. Introduce active listening by asking the group if they have had the experience of trying to talk with someone and not feeling heard. (Most group members, in our experience, relate to this experience.)

3. Let the group know that active listening is a way to talk about difficult issues. It is not necessarily a way to solve the problems, but it is a skill that can help people share their experiences.

4. Also point out that active listening can help us respond to a situation rather than react.

5. Divide the group into pairs (families stay together) and practice active listening: One person talks, the other listens and then repeats what s/he heard—without responding. When that is done successfully, switch roles, and the other person speaks. See worksheet for more detailed instructions on the activity.

6. Facilitate discussion of how difficult it is to listen without trying to convince the other person of your side.

7. Have group members complete the questions on the worksheet.

8. Ask how it feels to feel heard. Process experiences with this activity. Ask group members if they can think of times using this activity could be helpful for them. Ask if group members can see this as a tool to help them respond rather than react to situations.

GROUP DISCUSSION AND PROMPTS

- Most group members will be familiar with the experience of being "talked over" and/or interrupted. Group leaders can facilitate a demonstration in which a group member is asked to share something with the group leader, and the group leader interrupts and talks over the group member (not letting him or her finish). Then ask the group how they felt when this happened to the group member, as well as how they feel when this happens to them. Are they also guilty of this at times?

- Group leaders should be able to give examples of responding rather than reacting. For example, if a friend is late to a meeting time with you, an example of reacting would be yelling at your friend, whereas responding may be telling your friend how frustrated that made you feel and asking them to be more considerate. Responding is being able to step back from your emotions, whereas reacting is acting based primarily on your emotions.
- Presenting active listening as uncomfortable at first can be helpful.
- Group leaders should normalize difficulty with this activity, particularly the tendency to respond rather than just repeat, especially when the issues trigger strong emotions. Emphasis should be placed on the fact that eventually the listener will also have a chance to share his or her thoughts and feelings.
- Group leader can ask group members to identify situations in their lives in which active listening could be helpful.

AVOIDING PITFALLS

- Although active listening can decrease the intensity of difficult conversations, group leaders should encourage group members to first practice with topics that are not strongly emotional. Group leaders can also remind group members that it can be helpful to have a professional in the room when families attempt to first talk about volatile subjects.
- Group leaders should emphasize that active listening is not appropriate for couples to use at home if they have a history of violence.
- It can be difficult for group members to follow the rules of active listening. Group leaders should move from group to group as group members are practicing and be aware of interrupting, talking over, responding, and so forth.

MODIFICATIONS

- Group leaders could suggest that group members practice active listening with a non-family member at first and then move to working with a family member when they have some comfort with the practice.
- Group leaders can identify topics for group members to process. Some suggestions are:
  o Why am I in substance abuse treatment?
  o How do you make chicken soup?
  o Tell about someone you admire.
  o Tell about an event in your life that had a great impact on you.

CULTURAL CONSIDERATIONS

- This exercise can be particularly challenging for people who want to "fix" problems when they hear about them. Normalizing this experience can be helpful, as can starting a conversation with the group about cultural pressures to fix things and our tendency to want to do so. Many group members will support the experience of *not* wanting a solution to the problem but just wanting to be heard.

FOR GROUPS WITH YOUNG CHILDREN

- This activity can be facilitated in groups with young children. Group leaders can suggest topics such as: How I feel about doing chores (for children), or How I feel when you don't do your chores (for parents). As families become more familiar with active listening, it can become a good tool to help families talk about more difficult subjects, such as how the children feel about their parents' drinking or how they feel about their parents' fighting. For families that have not talked about these issues before, it can be helpful to begin to address them in family therapy sessions.

ACTIVE LISTENING

This is an exercise that can be useful when talking to someone about a sensitive or volatile issue, or any time you want to really be "heard." The point of the exercise is not necessarily to solve the problem but to truly listen to the other person, to hear what he or she is saying, and have the other person know that you heard him or her.

The Exercise

a. Choose one person to be the speaker and one to be the listener. The speaker chooses a topic. At first, choose a nonvolatile topic. Some examples are: What is your favorite movie and why? What was your favorite vacation and why? Talk about your family. Talk about what you are good at. Talk about what growing up was like for you. Talk about what is important to you.

b. The speaker begins to explain the topic to the listener. After a few sentences, the speaker stops talking, and the listener repeats what the speaker said. This does not have to be word for word, but it should be close enough that the speaker feels the listener has gotten the main points. If the speaker feels the listener has missed or misunderstood something, the speaker lets the listener know that and clarifies. The listener then repeats, and this process continues until the speaker feels s/he has been heard.

c. This process continues, with the speaker pausing after a few sentences and the listener repeating what s/he has heard.

d. Once the speaker is done, the partners switch roles, the new speaker speaks, and the new listener listens.

The Rules

DO NOT

- Interpret
- Answer or respond in any way other than repeating what the speaker says
- Try to solve problems

DO

- Listen
- Repeat and verify
- Notice how it feels (to be both the speaker and the listener)

What Was It Like...

To be the speaker? _____

_____

_____

_____

_____

To be the listener? _____

_____

_____

_____

_____

How would this be helpful talking with your partner about difficult issues? _____

_____

_____

_____

_____

6

Gratitude

Activity Title: Gratitude Quilt

Activity Mode: Expressive Arts (drawing)

RATIONALE

This is an expressive arts activity that gives the group an opportunity to acknowledge and express gratitude. In the cycles our groups experience, we at times experience a group that tends greatly towards negativity, blame, and resentment. We have found that this activity can help the group identify things they have to be grateful for while not negating the difficulties.

This exercise can be particularly helpful for family members that are experiencing a great deal of conflict. Again without negating the conflict, creating a gratitude quilt can bring in the realization that there are also things to be grateful for.

OBJECTIVES

- To help group members recognize they have things they are grateful for without negating difficulties
- To build appreciation
- To help group members recognize that it is possible to communicate more fully with family members and to help them see a way to do so
- To assist group members in offering support and encouragement to each other

MATERIALS

- Paper (white and/or colored, 8 ½ × 11-inch sheets)
- Markers, crayons, paints
- Tape
- Small pieces of paper (index card size), enough for each group members to have three pieces

OPENING QUESTION SUGGESTIONS

- What is something you feel grateful for today?
- What does gratitude mean to you?
- Who is a person in your life you are grateful for? Why?

METHOD

1. Introduce the activity with a discussion about gratitude. Open the floor to people's experiences.

2. Hand out three small pieces of paper to each group member.

3. Ask the group to each write three things they are grateful for (one on each small sheet of paper).

   a. One thing about their family they are grateful for

   b. One thing they are grateful to a family member for, but have not told them

   c. Anything else they are grateful for in their lives

4. Have each group member keep the second paper (b) and hand the other two sheets (a and c) back to the group leader.

5. Group leader mixes up the papers and then passes them out again (two to each member) so that the members do not hold their own slips.

6. Ask for a volunteer to begin reading, and have each group member read one slip. Go around the circle three times. Each person can decide when to read the slip that relates to her or his family member.

7. Once the slips have all been read, have the group draw/write their words and thoughts on 8 ½ × 11-inch sheets of paper—any way they want to. They can create one paper for each gratitude thought. They can draw the papers they currently are holding, or they can draw the gratitude thought they originally wrote.

8. If there is time pressure, group members can choose one gratitude thought to draw rather than drawing all three.

9. When people are finished drawing, have them tape the papers together in a "quilt."

10. Hang the quilt on the wall.

11. Instruct the group to take 1 minute of silence, go up to the quilt, and look at what they created. Ask them to pick one word or image that stands out to them. Then ask them to circle up (still in silence) and go around the circle, with each person sharing that word or image and why it touched her or him.

12. Ask the group to share experiences of what it is like to take some time and focus on things they are grateful for. Would it be helpful to do this every day? Group leaders can introduce the idea of a gratitude journal, in which the writer takes some time each day to write down (or draw) things he or she is grateful for. These do not have to be big things (probably lots of little things were identified on the gratitude quilt they just made). What does the group think of this idea?

AVOIDING PITFALLS

- For some groups, it can be helpful to have the drawing piece of this exercise done in silence also. This can be especially productive for groups that have a great deal of side chatter. Whether or not the drawing is done in silence, the final piece of the activity (having the group stand and look at the quilt and pick an image that stands out to them) is generally much more powerful if experienced in silence.
- Having the group draw all three images can take some time. Group leaders should be aware of the time and modify the activity (see below) if needed. Rushing to finish during the activity can take away from the power of the experience.
- Letting the group know that they can use words or drawings can be helpful in lessening anxiety that can come up when drawing is included in the activity.

MODIFICATIONS

- This activity is written with clients making three quilt pieces (one for each gratitude thought). If time is short, this can be modified so that each client creates one quilt piece—either with one gratitude thought or a piece that includes all three of their gratitude thoughts. They can use the ones they ended up with or the ones they originally created.

CULTURAL CONSIDERATIONS

- There are no particular cultural considerations for this activity.

FOR GROUPS WITH YOUNG CHILDREN

- This activity can be wonderful for groups with young children. Children's participation can often inspire the adult group members, both in terms of noticing things they are grateful for and in being more open to doing art.
- A family gratitude journal can be a fun and rewarding experience for families with young children, as well as a way to introduce a positive, daily ritual that all family members can take part in.

Resentment: Taking Back the Power

7

Activity Title: Letting Go Activity

*Activity Mode: Expressive Arts,
Experiential (drawing, ritual)*

RATIONALE

This activity was developed as a response to group members' request for an activity to help them with resentments they carry. In particular, after completing the Gratitude Quilt exercise, group members gave positive feedback about the exercise and also said that they would like an activity that directly addressed resentments.

This activity creates a space for group member to acknowledge resentments they carry and make a conscious decision about whether they are ready to let go of them. Having resentments is framed as normal and also as something that can impact us negatively (taking our power). Letting go of resentments is framed as an ongoing process rather than a one-time event.

This is not an activity for deeply processing resentments, and group leaders should direct group members to work with their individual therapists on those issues.

The activity also models a healthy letting-go ritual that clients can use in their lives.

OBJECTIVES

- To create a space in which group members can safely express resentments
- To acknowledge that having resentments is normal and to decrease the feeling that this makes group members "bad"
- To point out that we have some power over how much we allow resentments to take over our lives
- To provide an experiential exercise in which group members are encouraged to identify, express, and let go of resentments
- To frame working on decreasing the power of resentment as an ongoing process rather than a one-time event (similar to recovery)

MATERIALS

- Markers, crayons, paints
- Half-sheets of paper—enough for each group member to have three pieces. The exact size of the paper is unimportant.
- A metal can, safe for burning
- Matches or lighter

OPENING QUESTION SUGGESTIONS

- Do you tend to get angry and get over it, or do you hold on to anger?
- What do you think about apologizing?
- What is a resentment?
- How do resentments impact your life?
- What is the difference between holding a resentment and not trusting someone?

METHOD

1. Introduce the activity with a discussion about how resentments impact our lives. Open the floor to people's experiences.

2. Hand out three papers to each group member.

3. Ask the group to each write three resentments they are holding on to (one on each paper).

4. Ask for a volunteer to begin reading, and have each group member read one slip. Go around the circle three times.

5. Ask the group what it felt like to say the resentments out loud.

6. Once the slips have all been read, have the group draw and/or write about their resentments on the back of the papers. Group leaders can have other paper available if they need more room. Let the group know they can draw or write anything—there is no right or wrong.

7. When people are finished drawing, have them lay all the papers on the floor so all sheets can be seen.

8. Ask the group to stand in a circle around the resentments on the floor and take a moment of silence together.

9. Ask the group to talk about how holding these resentments impacts them. Do they agree that holding on to resentments takes your power?

10. Ask the group to gather their three resentments and think about whether they are ready to take some of the power back and release the resentment.

11. Let the group know they will be burning the resentments they are ready to let go of. If there are some they are not yet ready to let go of, they can either burn that resentment with the others or hold on to it and burn it later on, when they feel more ready to let go.

12. Go outside to a safe area, and inform the group that we will be burning the resentments to represent letting go, taking back our power. Though these experiences will always be a part of us, they do not always have to hold us so tightly.

13. Ask the group to do this part of the activity in silence, except the affirmation.

14. Light one resentment and let it burn in the fireproof can. Have group members put their resentments in the can one at a time until all are burned. As the resentments are put into the fire, the group member can repeat an affirmation, such as, "By burning this resentment, I take back my power."

15. When all the resentments are burned, group leaders can ask the group to again circle up and share experiences of the exercise.

GROUP DISCUSSION AND PROMPTS

- Ask the group how holding onto resentments impacts them. How do resentments take your power? How do they hold you back? Is there anything good that comes from holding resentments?
- What is the difference between holding a resentment and not trusting someone?
- It can be helpful to have a conversation about the usefulness of rituals such as this one. Can the group think of other rituals that could be helpful in this area?
- It can be important to emphasize that it may not be possible to completely let go of resentments and that people need not feel badly if this is the case. Group leaders should frame letting go of resentments as a process, and performing rituals such as these can help this process along.

AVOIDING PITFALLS

- One challenge for group leaders with this activity is to be aware of increasing negativity in the room and balance that in a way that results in a productive activity rather than a negative spiral. If group leaders are concerned the group is going in this direction, more of the exercise can be done in silence. Group leaders should emphasize the goal of the exercise is *not* to process deeply held resentments but to consider whether they are ready to let go of them. Some issues may need processing, and if this comes up during group, group leaders should direct group members to their individual therapists.
- Since this activity does involve burning the papers, care must be taken that this is done in a safe, contained manner.
- It can be challenging to keep the attention of the group focused on the activity once the group moves outside. To help with this, group leaders can remind the group before they go outside that they need to remain focused and silent for this part of the activity.
- In our experience, many group members will want to smoke when they go outside. Group leaders should decide whether they will allow this and set that rule before the group goes outside.
- It is important that family members not be rushed to let go of resentments they are holding regarding the addict's use or actions taken during active addiction. Group leaders should normalize these resentments, as well as the time it takes to be ready to let go. It can be helpful to have a conversation acknowledging those resentments, normalizing them, and asking whether the family member can imagine a time when they would not need to hold onto those resentments any longer. What would that time look like?

CULTURAL CONSIDERATIONS

- Since this exercise does involve some basic writing and reading skills, group leaders must be aware if there are group members that this is an issue for. Group leaders can work individually with those group members, being attentive to any embarrassment this may cause.

FOR GROUPS WITH YOUNG CHILDREN

- The concept of holding and letting go of resentments can be a useful concept for families with young children. Group leaders and families should consider whether this exercise could be helpful. Children, for example, may be holding resentments about their parents' drinking. These resentments should be processed in therapy before being addressed in this type of setting. It is important that children receive the message that having resentments is normal and natural, and they should not be rushed into letting go of those resentments.
- Group leaders may want to use another type of ritual rather than burning if working with children. An example would be ripping the resentments into very small pieces.

Curriculum Section VII:
Parenting

1

When Daddy Drinks

Activity Title: Functional Analysis

Activity Mode: Psycho-education

RATIONALE

This activity is a psycho-educational activity that addresses the fact that in families with substance abuse, all family members are impacted—not just the person abusing substances. We normalize the idea that family members often adopt unhealthy behaviors in order to cope with the substance abuse in the family. For example, Mom learns to protect Dad from the consequences of his drinking by calling in sick for him in the morning after a night of drinking. These behaviors are adopted as the individuals in the family seek to find ways of coping with the difficult situations that arise when substance abuse increases in the family.

This activity works to identify some specific patterns that have arisen in group members' families due to substance use. Once the patterns are identified, group members are encouraged to assess such behaviors and determine whether they are truly helpful. If the behaviors are not deemed helpful, group members are encouraged to consider whether they are ready to let go of the behavior, as well as identifying other behaviors they could adopt that would be more helpful.

These patterns and behaviors, even ones that are not helpful, are presented as coping behaviors that are developed as family members attempt to survive the impact substance abuse is having on their family.

Group members are also encouraged to consider patterns they learned in their families of origin that they may be ready to change. Again, these patterns are framed as survival skills that we all learn as children (not necessarily related to substance abuse, though many of our group members do grow up in that environment). The process of identifying, assessing, and ultimately choosing to either let go of or maintain those behaviors is presented as a normal, healthy part of life.

OBJECTIVES

- To help identify harmful patterns in group members' own and their family's behaviors and move toward more healthy ways of behaving
- To normalize behaving in unhealthy ways in an unhealthy situation
- To help group members see that people, particularly children, develop ways of behaving that are greatly influenced by what they see around them

- To develop hope that negative patterns can change
- To identify one "survival tactic" learned as a child that is no longer helpful and move toward being ready and able to let that go

MATERIALS

- Pens, pencils
- Functional Analysis worksheet

OPENING QUESTION SUGGESTIONS

- What is something you did as a result of someone else's substance use?
- What is something your family does to adapt to your substance use?
- What is something you remember learning? What was that like for you?

METHOD

1. Introduce the concept that in families, people behave in ways to try to make things work. Point out that as children, we learn to survive. Many of us learn to cope in ways that are not always healthy. Often we do things to survive as children and keep using those tactics as adults, even when we no longer need them. An example is that children growing up in dangerous households may learn that they can never ask for what they want. Many of these habits that made sense when we were children (in this example, it helped keep the children safe) are no longer necessary in adulthood. Even so, it can be challenging to let go of these behaviors and adopt other behaviors that are more effective.

2. Ask group members if they can think of behaviors of their own that may fit with this concept.

3. Introduce the concept that a family with addiction often functions around the addiction. Many families function quite well. Family members learn what role works best and fall into patterns and roles that feel very natural.

   Example: A child may be very quiet when Dad starts drinking and retreat to his/her room.

   Example: The entire family may know not to talk about anything important after Mom starts drinking.

   Ask group members for other examples from their own experiences.

4. Introduce the idea of doing a "functional analysis" of a family to help see what patterns come into play into their own families when people are using.

5. Pass out worksheet. Have each group member (families can work together, or each individual can complete a sheet and compare) decide who the "user" will be, and fill in the blank at the top of the page (What works when my _____ uses). This could be wife, husband, father, mother, son, daughter, aunt, uncle, and so forth—or it could be *I*.

Have each group member fill in the names and relationships of other family members (member 1, member 2, etc.). Then have them answer each question for each family member, as well as the remaining questions. The question reads: When _____ uses, Member 1 (write what member 1 does), and I (write what I do), because (write any reasons that come to mind—what is the function of this behavior?). To make this different, I could _____.

6. When all are finished with the worksheet, gather back together and facilitate a discussion about what the group learned about themselves and their families. Try to focus on what they found that they can do differently, as well as the survival skill they needed as a child but can now let go of.

AVOIDING PITFALLS

- It is important to be sensitive to the possibility that this exercise can bring up strong emotions and sometimes painful memories. Be sure to check in with the group about this, as well as emphasizing the importance of talking about any feelings of being triggered.
- This exercise can be confusing. Group leaders should be sure they go over the worksheet, and perhaps complete it themselves, to be sure they are comfortable describing the process to the group.
- The focus of this exercise is on changing one's own behaviors—not changing those of others. The Serenity Prayer (God, grant me the serenity to accept the things I cannot change, the courage to change the things I can change, and the wisdom to know the difference) can be a helpful tool to bring the focus back to changing oneself rather than other people.

MODIFICATIONS

- Rather than each individual completing a functional analysis, the family can work together to identify ways that it has adapted to substance use. This can trigger a useful conversation among family members, who often have very different individual experiences.
- Group leaders can create a family situation to use as an example for this exercise. See example worksheet.

CULTURAL CONSIDERATIONS

- Group leaders should be sensitive to culturally based structural differences of family functioning. In general, an interested and nonjudgmental stance is the most effective approach, which encourages group members to share their experiences. Group members will come from a variety of cultural and socio-economic backgrounds and will be more likely to share their experiences in an environment that feels supportive and nonjudgmental.

FOR GROUPS WITH YOUNG CHILDREN

- This can be an enlightening activity to facilitate with groups with young children. Parents will often be surprised at the insight children have into their parents' substance use. (We suggest that whatever parents think their children know about their substance use, multiply that by 10.) It can be very effective to have the children describe what they do in response to their parents' substance use. Parents should be coached to expect some surprises in this process and to be prepared for information from their children that they were not aware of. Group leaders should be aware of these situations and be available to give support to parents if this activity triggers strong emotions.

FUNCTIONAL ANALYSIS – EXAMPLE FOR GROUP LEADERS

What 'Works' When <u>*Dad*</u> *Uses?*

Example family

Dad drinks every day

Mom takes over everything when Dad drinks

2 daughters, me and Susie—Susie leaves the house or goes to her room to avoid Dad's drinking. I do everything I can to keep peace in the house.

Note: This example is written from the perspective of a child rather than an adult.

When Dad uses:

Member 1 (<u>Mom</u> **):** <u>looks after everything, gets mad</u>

And I <u>try to help Mom—I try not to make her mad</u>

because <u>if she gets mad, she and dad will fight and it is dangerous.</u>

To make this different, I could <u>talk to Mom when Dad isn't drunk and tell her how scary it is when she and Dad fight</u>.

Member 2 (<u>**Susie, my sister**</u> **):** <u>leaves the house, or goes to her room</u>.

And I <u>get mad at her and feel lonely</u>

because <u>when she's not there, I have to do her work, too. And I feel safer when she is there.</u>

To make this different, I could <u>ask Susie if I could go with her sometimes, and ask her to do her own chores. I could tell her I feel safer when she's here.</u>

What is one thing you can do differently in your family?

I can talk to my mom and my sister.

I want to talk to my dad, but I'm not sure I can.

I can spend more time with my sister.

I can ask my mom for help.

What is something you learned as a child that helped you survive but that you no longer need and are ready to let go of? What do you need to do to let that go?

This question is mainly written for adults, though children can have wonderful insight about things they can do differently as substance abuse plays less of a role in their family. In this example, the writer may decide to stop doing her sister's chores for her as she becomes more confident that the home is safe.

FUNCTIONAL ANALYSIS

What 'Works' When _____ Uses?

1. This exercise can be completed looking at the way things are in your family today or how they were when you were growing up.

2. Decide who will be the "using person" for this exercise. It can be any using family member—a parent, grandparent, other close relative, sibling, or child. Put this person's name in the blank above. The exercise can also be completed with yourself as the using person.

3. Write this same name in the first blank below.

4. Write the remaining family members' names below (Member 1, 2, etc.).

5. Answer the questions. The questions are the same for each family member, but the answers will be different depending on your relationship with each person. Think about

   a. What does Member 1 do when this person uses?

   b. What do you do?

   c. Why do you and Member 1 respond this way?

   d. Is this something you would like to change? If so, what could you do differently?

6. Complete this series of questions for other family members. For each family member, think about how this impacts your behaviors and if there are behaviors you would like to change. Note that this exercise is about changing your own behaviors, not that of others.

When _____ **(same name as in blank above) uses:**

Member 1 (_____): _____

_____

_____

_____

And I _____

_____

_____

_____

because _____

_____

_____

_____

To make this different, I could _____

_____

_____

_____

Member 2 (_____): _____

_____

_____

_____

And I _____

_____

_____

_____

because _____

_____

_____

_____

To make this different, I could _____

_____

_____

_____

Member 3 (_____): _____

_____

_____

_____

And I _____

_____

_____

_____

because _____

_____

_____

_____

To make this different, I could _____

_____

_____

_____

Member 4 (_____): _____

_____

_____

_____

And I _____

_____

_____

_____

because _____

_____

_____

_____

To make this different, I could _____

_____

_____

_____

Member 5 (_____): _____

_____

_____

_____

And I _____

_____

_____

_____

because _____

_____

_____

_____

To make this different, I could _____

_____

_____

_____

Member 6 (_____): _____

_____

_____

_____

And I _____

_____

_____

_____

because _____

_____

_____

_____

To make this different, I could _____

_____

_____

_____

What is one thing you can do differently in your family?

_____

_____

_____

_____

_____

What is something you learned as a child that helped you survive but that you no longer need and are ready to let go of? What do you need to do to let that go?

_____

_____

_____

_____

What About the Kids?

2

Activity Title: Parenting Styles—
How Can I Be A Better Parent?

Activity Mode: Psycho-educational;
Expressive Arts (acting)

RATIONALE

This activity works to help parents understand that there are different styles of parenting (Maccoby, 1992) and to identify parenting styles they use. We also work to relate the use of and comfort with particular parenting styles to the parenting styles of the families we were raised in.

When group members are presented with the various parenting styles, a discussion often comes up about the homes they were raised in as children. Many parents use parenting styles that are very similar to ones they were raised with. For example, we hear parents that experienced being spanked as a child defend spanking because it "worked for them."

On the other hand, we work with parents who make a conscious choice to do things differently with their children. For example, a parent who experienced physical violence as a child may avoid using physical punishment but may not see a problem with yelling and threatening behaviors.

One of the most common issues we see in families is the structural problem of parents wanting to be friends with their children and having a difficult time setting boundaries. Children in substance abusing families also are often "parentified" and take on adult roles in the family. A discussion of parenting styles can open the door to conversation about all these issues and more.

Most parents in our groups carry a weight of guilt and regret with respect to their children and not being good parents. Group leaders must be aware of and respectful of these feelings and experiences. In general, we find that acknowledging the pain of those feelings, as well as accepting that we can't change the past (but can change what we do from here on), is most effective.

Not all our group members are parents. In mixed groups such as these, group leaders must be adept at having this exercise be meaningful to both parents and nonparents in the group. This can be accomplished by facilitating a discussion that includes both parents and nonparents and flows between focus on family-of-origin experiences and current experiences of parents.

OBJECTIVES

- To introduce the concept of different parenting styles
- To introduce four different parenting styles and help group members identify the style of parenting they were raised with and (for those that are parents currently) how that relates to the parenting style they use today

- To help parents identify areas of parenting they would like to modify and develop a direction they want to go in
- To normalize feelings of sadness and regret regarding children and parenting and create hope for a better future for themselves and their children

MATERIALS

- Paper, pens
- Parenting Styles worksheet

OPENING QUESTION SUGGESTIONS

- What is one thing that you remember from your childhood that you would like to have your children experience?
- What is a hope you have for your children? If you don't have children, what is a hope you have for yourself?

METHOD

1. Introduce the idea of parenting styles and ask the group to share what it was like to grow up in their families. Introduce the Parenting Styles worksheet.

2. Review the four different parenting styles and ask for examples from the group. Have the group follow the worksheet and mark the identifiers that most closely describe their own parents. **For members that are currently parents:** Have them also mark the descriptors that most closely identify their own parenting style. This can lead to a discussion of how the parenting we experienced as children impacts how we parent.

3. Point out that no parent uses one style 100% of the time; all parents will probably use some of each style at one point or another. We will, however, tend to favor one style or the other.

4. Give time for the group to identify which parenting style is most comfortable for them.

5. Give time for the group to discuss whether they would like to move toward another style of parenting. Encourage them to identify three things they want to do differently as parents.

6. Ask for volunteers to choose a parenting style to act out one of the role plays. Each role play will be acted four times (one for each style).

7. Gather the group together and ask what they learned from this activity.

GROUP DISCUSSION AND PROMPTS

- Be sure to emphasize that there is no perfect parent and that as parents, we all make mistakes and look back and wish we had done things differently. Talk about the fact that though we cannot change the past, we can make the future different by making changes, beginning right now.

- Emphasize that every parent can be a better parent. Point out that we are not born knowing how to be parents, and many of us do not grow up with good examples of wonderful parenting skills. Parenting skills are something that we can all improve on. Taking a parenting class one time does not mean that we are finished working on being better parents!

- Most parents in our groups carry a weight of guilt and regret with respect to their children and not being good parents. Group leaders must be aware of and respectful of these feelings and experiences. In general, we find that acknowledging the pain of those feelings, as well as accepting that we can't change the past (but can change what we do from here on), is most effective.

- It can be helpful to introduce a discussion about how we learn to be parents. Group members may talk about learning to be parents by having role models that they follow, whether in their own family or in other families. Some group members may also talk about never having learned to be good parents because they never experienced good parenting themselves. This can be a powerful topic of discussion for the group.

AVOIDING PITFALLS

- This activity, like all activities that relate to parenting and children, can stir strong emotions in group members. Group leaders must be aware of these situations and offer a supportive, nonjudgmental environment in which parents in the group feel safe to talk about their experiences and feelings. Normalizing these experiences can always be helpful. Group leaders must work to find a balance of normalizing difficult experiences while at the same time holding the idea that every child deserves to be raised in a healthy, nurturing environment. The fact that the parent is in this group and working toward improving his or her life and maintaining sobriety is always a way to point out positive changes and a direction for hope.

- Some group members may stand strongly for spanking. Although the laws about spanking vary from state to state, we generally suggest that it is much more effective long-term for parents to identify consequences other than spanking or "whooping." There is a great deal of information available on this choice, and group leaders may suggest that parents look into this information. However, it never helps for group leaders to get into arguments with group members! In such situations, group leaders can model "agreeing to disagree."

MODIFICATIONS

- The group can be divided into four smaller groups, with each group taking on one style of parenting for each vignette. Groups can expand on the situations as they would like (and as appropriate).

- Each group can take one vignette and act it in all four different parenting styles.

CULTURAL CONSIDERATIONS

- The main issue that comes up in this group is related to spanking. In our area, spanking or "whooping" is very much the norm, and many of our group members strongly believe that spanking is a necessary part of childrearing. We suggest that group leaders have clear understanding of the laws of the particular area they are working in and pass that information on to group members. Group leaders can also help individuals gather information about the long-term effects of spanking children if group members are interested. This can be a challenging issue to process in group, as many individuals (group leaders included) have strong feelings, both pro and con, about physical punishment for children. In general, we believe that it is most helpful if group leaders are honest about their biases and continue to be respectful in all situations.

- In most states, group leaders are required to report to Department of Social Services if there are concerns about safety for children. It is important to be familiar with the requirements in your area. This is something we let our group members know about up front. If it becomes necessary to make a report, we generally tell the group member of the requirements and offer him or her the opportunity to make the report him- or herself or with us. This is not the case if there is a concern for safety.

FOR GROUPS WITH YOUNG CHILDREN

- This activity is more appropriate for groups without young children.

PARENTING STYLES

Read the following descriptions of four different parenting styles, and check the descriptions that match how your parents parented you. Remember that all parents have a blend of different styles rather than only one.

Authoritative

- ❐ Parent is demanding, but responsive to the child
- ❐ "Tough love"
- ❐ Wants child to be increasingly independent, but within a set of rules and boundaries
- ❐ Parent is warm and nurturing
- ❐ Allows the child freedom to explore, but also to learn from mistakes and take the consequences for his or her actions
- ❐ Clear standards for behavior are set, explanations given
- ❐ Open communication between parent and child is encouraged
- ❐ Other: _____

Authoritarian

- ❐ Parent is demanding, but is not responsive to the child's point of view
- ❐ Strict style of parenting
- ❐ There is little open communication between the parent and child
- ❐ There are rules and boundaries, but the child may not be told the reason behind them
- ❐ Other: _____

Permissive

- ❐ Parent is not demanding but is responsive to the child's point of view
- ❐ Parent is very nurturing and warm
- ❐ No demands are made on the child
- ❐ Parent makes very few rules or little structure for the child
- ❐ Children are allowed to act freely and usually get their own way
- ❐ Children raised in this style tend to engage in more risky behaviors, such as drugs and alcohol, when they enter their teens
- ❐ Other: _____

Neglectful

- ❐ Neglectful parents are detached and uninvolved
- ❐ Parent is not demanding and not responsive
- ❐ Parent doesn't give child emotional support
- ❐ Limits are not set for the child
- ❐ Child feels that he or she is not an important part of the family
- ❐ Parent will provide food, shelter, and other basic needs but is not involved otherwise
- ❐ Children raised in neglectful households may be withdrawn socially as adults. They may become involved in drugs or alcohol to help them control shyness, anxiety, or other mental health issues.
- ❐ Other: _____

In My Family

My mother's parenting style was mostly _____.

My father's parenting style was mostly _____.

My grandparents' parenting style was mostly _____.

My own parenting style is mostly _____.

I would like to include some aspects of the _____ parenting style in my own parenting.

Some things I would like to do differently are:

_____

_____

_____

_____

Role Plays

- Child did not do chores as requested
- Child doesn't want to go to bed
- Child refuses to eat dinner
- Child will not pick up toys

What Do We Tell the Kids?

Activity Title: What Do We Tell the Kids?
(Boundaries, Structure)

Activity Mode: Psycho-educational

RATIONALE

One of the most common difficulties we see in families with substance abuse is a distorted family structure. In more healthy-functioning families, parents take on a parental role, and children are able to function as children. In families with substance abuse, parents often behave more like children than adults, resulting in a lack of structure, parental responsibility, distinct roles, healthy and regular routines, and the like. Feeling unsafe, children often behave in "parentified" ways, worrying about adult issues such as money, food, safety, and so on.

Once parents begin on the path toward sobriety, they often are left with a great deal of guilt about how their substance use has impacted their children. We see many parents in our groups struggle with these feelings. Often, we see parents try to make up for this impact on their children by giving in to their children's demands and having difficulty setting boundaries. We try to help parents see long-term benefit from structure and boundaries for their children, as opposed to the short-term gratification of giving them what they want in the moment. This is something we talk with our parents about regularly.

We also work with parents who have difficulty being a parent as opposed to being a friend. We try to help our parents recognize that their acting as parents will be much more beneficial for their children's long-term growth than will being the children's friend. Helping parents recognize this difference and move toward setting appropriate boundaries, thereby strengthening the healthy structure of the family, can have a powerfully positive impact on families.

OBJECTIVES

- To help parents conceptualize family structure, particularly the importance of maintaining healthy boundaries between the parent's role and children's role
- To help parents think about and distinguish between information that is appropriate to share with children and information that is not

227

- To normalize the fact that when families have addiction, the family roles often became disorganized, and when families move into recovery, it is necessary to repair this
- To acknowledge the difficult of setting clear boundaries for children and following through
- To acknowledge the pain of knowing parents have let their children down
- To build hope in the idea that parents can be good parents from now on

MATERIALS

- Pens, paper
- Worksheet: What Do We Tell the Kids?

OPENING QUESTION SUGGESTIONS

- What is the difference between being a parent and being a friend?
- What is an example of something you did *not* get when you wanted it but that ended up being a good thing (or something you *got* when you wanted it, but it ended up being not good).

METHOD

1. Open a discussion about parental roles and childrens' roles in the family. Emphasize the importance of having these roles be separate, and point out that in many families (particularly in those with substance abuse), these roles become confused—parents act like children and children act like parents ("parentified"). Normalize this while at the same time emphasizing the importance (and possibility) of changing it.

2. Have parents complete the diagram of their family structure and complete questions related to this diagram. Have them draw circles to represent each family member and place them on the diagram based on their function (who acts like parents, who acts like children).

3. Before completing the remainder of the worksheet, facilitate a discussion of the activity up to this point. Many people in the group will be familiar with the concept (if not the term) of parentified children (many were forced to act like adults in the families they grew up in!). Encourage discussion.

4. When appropriate, move on to the next questions. The group can take time to write their answers, or if the discussion is lively, it is often helpful to have group members answer these questions verbally. Be sure to include nonparents in the discussion by asking for their experiences as children themselves.

GROUP DISCUSSION AND PROMPTS

- Point out that struggles with these issues are very normal for parents, and there is no black-and-white answer for most of them. The important points are that children need consistency, love, support, safety, and to know their parents are there for them.

- Also point out that these issues come up in all families, not just in those with substance abuse. Learning about parenting should be an ongoing process. Since parenting requirements change as children change, there is no time when parents should be done with the learning process.
- It can be helpful for group leaders to be familiar with parenting skills classes offered in the community and encourage group members to attend. Group leaders may respond by saying they have already taken parenting classes, in which case the group leader can again refer to the previous point.
- Parents' feelings of guilt can have a profound impact on their willingness to set boundaries. This topic should be addressed directly in this group.

AVOIDING PITFALLS

- As with all groups related to parenting (as well as other issues), group members may have strong feelings arise. Acknowledging these feelings and offering support and encouragement to parents are essential.
- Group members will sometimes have very different ideas about parenting and may react strongly to some of the ideas presented here. For example, we have had group members who were absolutely committed to being their child's best friend and to telling her everything. In these cases, group leaders need to continue to present the information and avoid falling into an argument with any group members. It can be helpful to turn the issue to the group and get feedback from other group members.

MODIFICATIONS

- Group leaders can have the group act out skits related to this topic. For example, one skit could be with a parent sharing fears about their financial situation with a 7-year-old child, while the other could be reassuring the child and reminding him that the parent will look after everything and he doesn't have to worry. Group leaders can identify several skits, divide the group into teams, and have them each act out a skit. This can be a good hands-on way for group members to process several different issues.
- Even better: Group leaders can ask the group to come up with issues, and small groups can act out skits using those ideas.

CULTURAL CONSIDERATIONS

- There will be cultural differences in the amount of involvement children have in adult affairs. In general, we suggest that these differences be acknowledged and processed. Group leaders may not be aware of these differences and must maintain an open curiosity to these issues. Group leaders can also help individuals process the impact of these differences on children and help group members identify ways to develop the most healthy environment possible for their children.

FOR GROUPS WITH YOUNG CHILDREN

- Since this group is about parenting issues, it is generally not suitable for young children. By clarifying this, group leaders are modeling discriminating between situations helpful for children and those that are not and setting and holding that boundary.
- Group leaders can help parents identify ways of presenting the new boundaries they are learning to children. For example, if a parent is not going to share financial issues with the child that he or she has shared in the past, the child may be confused and/or upset. Group leaders can coach the parent in how to respond. In this case, the parent could say, "I know in the past I have shared that with you, but I realize that's an adult issue, and you don't have to worry about it. I will look after things, and we will be fine. I love you." In this way, the parent is setting a clear boundary and, at the same time, letting the child know that everything will be fine.

WHAT DO WE TELL THE KIDS?

Parental Role

Parent–
Child _____
Boundary

Child Role

Instructions: Place one circle to represent each family member on the above diagram. Think about where to put each person based on her or his behavior and role in the family. People in more of a parental role (or children that are involved in parental decisions and/or information) should be placed above the parent–child boundary. Some people may be right on the line. Parents who take less parental responsibility in the family should be placed below the line. In many families, parents and children can be placed in equal places. The resulting diagram will represent how the family functions in terms of parent/child roles.

What I Learned From This Exercise

(Think about placement of parents relative to children. Do children need to act more like children, or do parents need to be more parent-like? What are some specific changes you can make?

_____

_____

_____

_____

How has substance abuse impacted the structure of your family?

_____

_____

_____

How Much to Tell the Kids

Often, parents who have had problems with substance abuse feel guilty and want to make it up to the children. Though this is understandable, sometimes parents confuse being good parents with giving children what they want. Setting and maintaining boundaries and limits are important parts of parenting and actually help children feel safe and secure—even though in the moment, they will cry, kick, scream, and do/say anything to get their way. Parents, feeling guilty about their own substance use, often find it difficult to maintain these boundaries and limits in the face of the children's unhappiness in the moment.

What is the difference between being a friend and being a parent? How do you manage this with your children? What do you struggle with?

_____

_____

_____

_____

Parents will sometimes say that their child is their best friend. How do you think this would impact a child?

_____

_____

_____

_____

As a result of this exercise, do you recognize any structural changes in your family that need to take place in order to create a safer, more stable home for your children? What changes? What do you need to do to make those changes?

_____

_____

_____

_____

How will you answer your child if s/he asks for information about parents' issues, such as money, housing, substance abuse, and so forth?

_____

_____

_____

Consequence With Empathy

Activity Title:
How Can I Be a Better Parent?

Activity Mode:
Psycho-education, Role Play

RATIONALE

Many parents who come to our groups feel at a loss when it comes to knowing how to set boundaries and respond to their children in ways that are productive. As parents learn to take on the role of parent in the family, it can be helpful and inspiring for them to have some concrete parenting tools.

Giving consequences with empathy is a main tool of the Love & Logic parenting program (Cline & Fay, 2006) and can be one of the most effective and useful tools a parent can learn. When parents learn to give a consequence with empathy, they:

1. Allow their child to experience the consequences of his or her behaviors so the child learns that his/her actions have consequences

2. Encourage the child to be frustrated with his or her own behaviors rather than with the parent

3. By responding with empathy, the parent actually experiences empathy for the child, decreasing her or his own frustration and irritation and interrupting the escalation of emotions

Parents, particularly parents that carry a weight of guilt about how they have parented their children, often need help in recognizing the importance of having their children experience the consequences of their behaviors. It is much less painful for children to learn that actions have consequences when they are young and the consequences are bearable. A 6-year-old experiencing a direct consequence of hitting his mother (perhaps his mother will not want to play with him for a while) will learn that hitting will not get him what he wants. A child who does not learn this may be forced to learn it as an adult, when the consequences may result in much more severe consequences, such as jail.

OBJECTIVES

- To remind parents there is no such thing as a perfect parent—we all make mistakes
- To remind parents that we can all become better parents and help parents identify three things they can change to do so
- To learn a concrete skill (consequence with empathy) to make parenting easier
- To help parents recognize the importance of allowing their children to experience the consequences of their actions
- To have each parent pick her or his own empathy statement and practice using it
- To learn and practice the skill of giving consequences with an empathy statement

MATERIALS

- Paper, pens
- How I Can Be a Better Parent worksheet

OPENING QUESTION SUGGESTIONS

- Can you think of a time your parents shielded you from the consequences of your actions (or you shielded your children)? What was the result?
- Can you think of a mistake you made that you learned from?
- For parents—can you remember a time when you wanted to step in and save your child from consequences, and you didn't? What was that like?

METHOD

1. Hand out worksheet, How I Can Be a Better Parent. This worksheet is in two sections:

    a. The first section allows group members to identify three areas in which they want to do things differently as a parent. Emphasize to the group that this is not about being a bad parent, but improving parenting skills is hopefully something parents do over the course of their whole lives (once a parent, always a parent). Give the group time to identify three changes they would like to make and ask them to share what the changes are and how they can make the changes. Remind them to be specific and that the changes do not have to be big. Give them some varied examples (I will take three breaths before I yell; I will take a time out; I will sit down with my children for dinner; I will play with my child for 15 minutes; I will read to my child before bed, etc.).

    b. The second section is a variation of a technique taught through the Love & Logic parenting program: giving a consequence with empathy. This technique can be a helpful tool that parents can take from the group and use frequently with their children. One of the keys to success with this technique is **practice**. Another key to success is **being genuine** with the empathy. Children will react negatively to sarcasm. Make sure you have enough time for the group to split into pairs and practice this technique. The group leaders should be able to give numerous examples of empathy statements with a consequence, such as the following:

   i. Oh, that's sad, your homework's not done, so you won't be able to watch TV tonight. Maybe tomorrow you'll get it done in time.

   ii. Oh, darn, you ran up the slide again. That's really sad; now you can't play on the slide any more today.

   iii. Oh, what a drag, your toys are still on the floor. You won't be able to play outside now.

   iv. Ooooooooh, that's too bad. You pushed Joey, so now we have to go home. I was really hoping you would be able to play with Joey today.

2. **For Group Members That Are Not Currently Parents:** Have nonparent group members pair up with members that are parents, and have the parents practice the empathy/consequence statements with the nonparent member. Ask the nonparent to give the parent feedback as to whether the statement felt genuine.

3. Gather the group together, and ask members to share the experience of using the empathy/consequence statements and to share the statement they developed that seemed most comfortable for them. Ask whether this technique seems useful for them.

GROUP DISCUSSION AND PROMPTS

- It often takes practice to become comfortable with an empathy statement. If the statement is said with sarcasm, it will not work! If parents are not able to be genuine with the empathy, they should keep practicing before they try to use this tool with their children. Do not be sarcastic!
- When parents begin to set boundaries, their children's behavior will often get worse before it gets better. This does not mean that the parenting is not working—in fact, it is a good sign. The key is for parents to be consistent and not give up. With time, children will see that their parents are serious, and things will get better.
- Encourage parents to identify an empathy statement that feels comfortable for them; they are not limited to our examples.
- Saying the empathy statement out loud can actually help the parent stay calm and avoid falling into an argument with the child. If a parent is arguing, she or he has already lost! Ask the group who has had this experience (noticing themselves arguing with their 4-year-old)?
- The idea of learning parenting as we go along can be a helpful idea to combat the common notion that anyone who needs to take a parenting class must be a bad parent. Group leaders can normalize the experience of struggling with parenting and needing help, support, and encouragement—and new ideas. Our children are always changing, and as parents, we need to continue to change as well. We are not born knowing how to parent, and many of us did not have very good role models.

AVOIDING PITFALLS

- The most common pitfall with this exercise is sarcasm. If parents use this tool with sarcasm, it can do more harm than good. Group leaders should be quick to point out tinges of sarcasm if they notice it, and encourage parents to be honest with themselves about what they are feeling and *not* use this technique if they are using sarcasm.

- Normalizing this as a simple tool, but one that can be challenging to learn at first, can be helpful. Having a number of examples of behaviorally challenging situations so the parents can practice giving a consequence with empathy numerous times can be useful. Parents can bring their own struggles with their children, and the group can give ideas how this tool can be used in the various situations.
- Taking time to help the group to acknowledge the importance of allowing children to experience consequences can be helpful in the overall process.
- Group members may bring up the idea of spanking as a consequence. This can be a sensitive issue, and in many cultures, spanking is an accepted consequence. When spanked, a child often blames the parent for the spanking rather than taking responsibility for his/her own behaviors. The long-term goal of the child receiving consequences is to help the child build a sense of responsibility. Being able to blame the parent can hinder this process.

MODIFICATIONS

- Group members can participate actively in developing scenarios in which a child misbehaves. The group can role play the scenarios using empathy, followed by a consequence.
- Group members can role play scenarios twice, once with great sarcasm, once with very little. This exercise can help emphasize the importance of not being sarcastic. These skits can be quite humorous.
- It can be helpful to practice this activity over a 2-week period. At the end of the first session, group leaders should give homework to practice giving consequences with empathy during the week, and come back the next week to talk about how it worked. The second week's group can be used to continue practicing different scenarios, using the actual experiences of families over the week.

CULTURAL CONSIDERATIONS

- As mentioned above, spanking can come up as a possible consequence. Different cultures have different views on spanking, and many of our group members are committed to using spanking in their families. We try to respect this perspective, as well as encourage parents to develop a variety of approaches to problem solving. We emphasize the benefit of not being tied to one response and of building a toolbox that has tools for many situations.
- Cultural differences in children's relationships with their parents and how children are expected to behave in families can come up in any discussion around parenting and childrearing. Our approach is that it is not necessary to have the same view of how families should be structured or how they should function. It can always be helpful, however, for parents to be able to respond in a variety of different ways in a variety of situations.

FOR GROUPS WITH YOUNG CHILDREN

- This activity is meant to be facilitated with parents and for parents to practice giving consequences with empathy at home.

HOW I CAN BE A BETTER PARENT

- There is no such thing as a perfect parent
- All parents make mistakes.
- All parents look back and wish they had done some things differently.
- We can all, always, become better parents.

Three things I can do differently in parenting my children:

1. _____

   _____

2. _____

   _____

3. _____

   _____

Consequence With Empathy

- Boundaries and limits help children feel safe
- Children need to experience the consequences of their behavior while the consequences are survivable
  - Empathy helps the child blame the action, not the parent
  - Empathy helps the parent remain calm
  - Choose your empathy statement (That's so sad, oh darn, bummer, Ooooooh, etc.) and practice!
- Empathy statement first...then consequence
  - Oh, that's so sad, you threw the stick at Johnny, now you can't play with him until tomorrow.
  - Oooooohhh, darn, you didn't pick up your toys, now you won't be able to play with them any more today.
  - Oh, bummer, you forgot to take the trash out and I had to take it out after you went to school. Now you'll have to fold the wash for me. I know that takes more time. Darn, I really hope you'll remember the trash tomorrow.

Remember: This takes practice!!! It does not feel comfortable at first!

The empathy statement that seems best to me is:

_____

Or

_____

Now think of three times in the past week that you could have used an empathy statement (with a consequence) and write down what you could have said:

1. _____

2. _____

3. _____

Now check:

- Does each statement have an empathy statement?
- Does each statement have a realistic consequence?

Bibliography

Atkins, S., Adams, M., Chester, T., McKinney, C., McKinney, H., Rose, L., Wentworth, J. (2000). *Collective voices in expressive arts*. Boone, NC: Appalachian State University.

Black, C. (1987). *It will never happen to me*. New York: Ballantine Books.

Campbell, S. (1980). *The couple's journey*. San Luis Obispo, CA: Impact Publishers, Inc.

Center for Substance Abuse Treatment. (2006). *Client's handbook: Matrix intensive outpatient treatment for people with stimulant use Disorders*. Rockville, MD: HHS Publication No. (SMA) 09-4154.

Cline, F., & Fay, J. (2006). *Parenting with love and logic*. Colorado Springs, CO: Navpress Publishing.

Edwards, J. (2003). *Working with families*. Durham, NC: Foundation Place Publishing.

Felder, R., & Weiss, A. (1991). *Experiential therapy: A symphony of selves*. London, MD: University Press of America.

Maccoby, E. E. (1992). The role of parents in the socialization of children: An historical overview. *Developmental Psychology, 28,* 1006–1017.

Maccoby, E. E., & Martin, J. A. (1983). Socialization in the context of the family: Parent–child interaction. In P. H. Mussen & E. M. Hetherington, *Handbook of child psychology: Vol. 4. Socialization, personality, and social development (4th ed.)*. New York: Wiley.

McGoldrick, M., & Gerson, R. (1986). *Genograms in family assessment*. New York: W. W. Norton.

Mead, G. H. (1934). *Mind, self and society*. Chicago: University of Chicago Press.

Merton, R. K. (1949). *Social theory and social structure*. New York: Free Press.

Minuchin, S. (1974). *Families and family therapy*. Cambridge, MA: Harvard University Press.

Nichols, M., & Schwartz, R. (2001). *Family therapy, concepts and methods*. Needham Heights, MA: Allyn & Bacon.

Pearson, C. L. (1993). *Women I Have Known and Been*. Washington, DC: Gold Leaf Press.

Rawson, R., Obert, J., McCann, M., & Ling, W. (2005). *The matrix model, family education group handouts*. Center City, MN: Hazelden.

Satir, V. (1983). *Conjoint family therapy*. Palo Alto, CA: Science and Behavior Books.

Thrower, S. M. et al. (1983). The family circle method for integrating family systems concepts in family medicine, *Journal of Family Practice, 15* , 451–457.

Watzlawack, P., Weakland, J. H., & Fisch, R. (1974). *Change: Principles of problem formation and problem resolution*. New York, NY: W. W. Norton.

Wegscheider, S. (1981). *Another chance: Hope and health for the alcoholic family*. Palo Alto, CA: Science and Behavior Books.

Winek, J. L. (2010). *Systemic family therapy: From theory to practice.* Thousand Oaks, CA: Sage.

Winek, J. L., Dome, L. J., Gardner, J. R., Sackett, C. R., Zimmerman, M. J., & Davis, M. K. (2010). Support network intervention team: A key component of a comprehensive approach to family-based substance abuse treatment. *Journal of Groups in Addiction & Recovery, 5*, 49–60.

www.FamilyTies.org, *Genogram symbols.*

About the Authors

Joan Zimmerman, MA, LMFT, LCAS, has been working with substance abuse and families for nine years. Until recently, she supervised the Family Solutions substance abuse treatment program. This program was specifically designed to work with families with substance abuse. She was involved in initially designing the program, with broad input from a variety of community members and agencies, and has consistently led weekly Family Groups for the families the program works with. Currently, she is in private practice and teaches in the Human and Psychological Counseling department at Appalachian State University. She is a licensed Marriage and Family Therapist, a licensed Clinical Addiction Specialist, and an approved Marriage and Family Therapy supervisor.

Jon L. Winek, PhD, LMFT, LPC, currently directs the Marriage and Family Therapy program at Appalachian State University, where he has been teaching since 1993. He has been involved with the theoretical development of the treatment program at Family Solutions, and continues to provide ongoing clinical supervision to the staff. He is the author of the DVD series *Systemic Family Therapy: From Theory to Practice*, as well numerous book chapters and journal articles.

⑤SAGE research**methods**

The essential online tool for researchers from the world's leading methods publisher

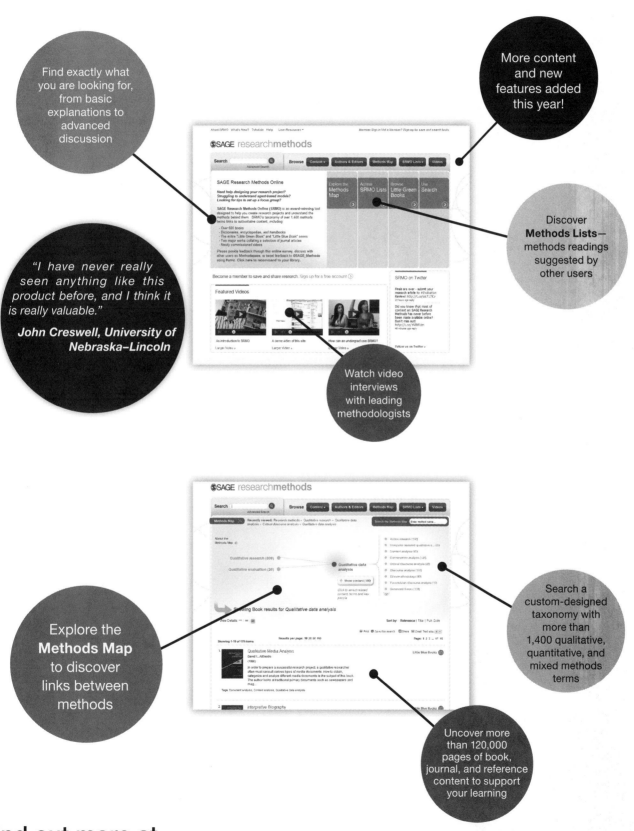

Find exactly what you are looking for, from basic explanations to advanced discussion

More content and new features added this year!

"*I have never really seen anything like this product before, and I think it is really valuable.*"

John Creswell, University of Nebraska–Lincoln

Discover **Methods Lists**— methods readings suggested by other users

Watch video interviews with leading methodologists

Explore the **Methods Map** to discover links between methods

Search a custom-designed taxonomy with more than 1,400 qualitative, quantitative, and mixed methods terms

Uncover more than 120,000 pages of book, journal, and reference content to support your learning

Find out more at
www.sageresearchmethods.com